Tulane Studies in Philosophy

Volume VII

THE NATURE OF THE PHILOSOPHICAL
ENTERPRISE

ISBN-13: 978-90-247-0281-7   e-ISBN-13: 978-94-011-7638-5
DOI: 10.1007/978-94-011-7638-5

TABLE OF CONTENTS

THE SUBJECT-MATTER OF PHILOSOPHY

Edward G. Ballard

THE expressions used so often to define philosophy, — e. g. philosophy is the study of reality; it is the attempt to understand the whole of our experience; it is the daughter of religion and the mother of the sciences,—require a philosophy in order to interpret them sensibly. I should like to maintain the view that philosophy is the continuously renewed interpretation of a certain subject matter,[1] but I am aware that this description, too, is ambiguous, since whatever it is that philosophy interprets and however it proceeds to make the interpretation have always appeared as changing questions. In a brief paper all the problems relating to philosophic interpretation can scarcely be broached. I shall limit myself to stating as well as I can what the subject-matter is which philosophy uses. In so doing I shall of course be using a technique and a doctrine of interpretation. Still it will be no part of the purpose of this paper to make this doctrine and technique explicit. It will be enough for the present if the datum of philosophy be adumbrated.

What, then, is the datum which philosophic effort attempts to interpret? One time-honored answer to this question is that the subject matter of philosophizing is all that is. Such an answer as this is so general and vague that it leaves us with an equivalent question on our lips. Other answers to this question move along

---

1 No doubt philosophy has always been regarded in some sense as interpretation. Some philosophical interpretations, however, are supposed to be final, either independent constructions or systems, said to include and explain all the sciences and arts, even all of reality, or else constructions dependent upon some favored science or art. In suggesting that philosophy is a *continuously renewed* interpretation, I mean to separate the present view from the belief that philosophy is or possibly can be any such complete system. Philosophy, as it has been recorded, is a continuously changing translation and interpretation of a set of problems related to a certain subject matter. The person who believes that he possesses the definitive solution to a philosophic problem is, doubtless, no less in error than the person who believes philosophic problems to be altogether irrelevant to his concern.

the possible alternatives, increasing in precision and also in narrowness, until a positivist's answer is reached; philosophy considers the language which philosophers and scientists are currently using. Rather than attempt to criticize such answers as these, I shall offer one more. Philisophy, I suggest, comes into being as the consequence of interpreting archaic experience.

What, then, is archaic experience? The term 'experience' is used here in a broad sense to indicate any object or process of concern. For example, among the more obvious sorts of things which elicit concern are these: the self, the world, and, if we are to persist, the concern for survival and the arts useful thereto. 'Archaic' may be used in both a logical and a temporal sense. Temporally its reference is to the experience of early man as he confronted his cravings and the world around him. Since we are to determine what the primary datum of subject-matter of the western tradition in philosophy may be, experience which is archaic in the temporal sense relative to it must be that which is to be defined.

Now it is very likely, I assume, that uninterpreted experience is to be found nowhere; at least since Kant wrote, this assumption has been widely accepted, and it will not be defended here. If we accept it, then our datum, no matter where we begin, will always be expected to contain an element of interpretation. In philosophy as elsewhere, prime matter is probably undefinable. Let us choose, as the most nearly primary datum which we can find, that point in history where the content and the meaning which define the human concern can be seen to stand out simply and at the same time to contain those elements which are significantly developed in later time. Thus the experience which is archaic relative to this inqury is that which presents in simplest available form the subject matter which later analysis and construction will interpret. An interpretation may in turn render a more sophisticated experience possible which *might* then become the subject matter of finer and more fertile interpretation,—and so on. Our topic will confine us mostly to the earliest preserved stage of this process. I shall examine briefly two or three views of the content of early Greek thought, discuss a widespread antique interpretation of it, and then try to isolate certain elements which became central in Plato's dialogues and thereupon in the remainder of the western philosophic tradition.

Textbooks on the history of phisosophy have made up passingly familiar with early questions and with supposed answers about man and the universe. Philosophy begins in wonder, it is said, and the first object of wonder was the nature of the world. The Milesian answers to such questions about the nature of nature are supposed to be more or less exhausted by references to water or air or the unlimited which are thought to have set their primitive wonder at rest, at least for a time. And yet there is a place for skepticism. These answers, or at least the way in which they are commonly presented, seem to be at once more sophisticated and more náive than one would have expected from a Greek of the sixth century B. C. All things are made of water. "How almost right!" one may exclaim, especially if one happens to be thinking of a watermelon. And even quartz, if heated, gives off its water of crystalization. Can this be, though, the way in which Thales thought? Or can this be the kind of problem which he thought about? I think not. It is easy for us to read our culture and its problems into the early Greeks. At first glance our language seems to fit. At second glance, though, it obviously does not. Thales also said, "All things are full of soul (ψυχή)". Pheredydes, a contemporary, said that from a primal seed (γόνος) was born fire, air, and water. These remarks do not sound like the observations of clever but uninstructed naturalists. They have a primitive ring about them which bring us to suspect the common-sense intention and meaning behind the supposed empirical generalization that all is water.

An entirely different way of getting at the original meaning of these early philosophical dicta is presented by F. M. Cornford in *From Religion to Philosophy*.[2] His method is to interpret the early philosophical fragments in terms of notions known to be common among the Greeks from a study of their own poetry, religion, and especially their social organization. Thus he attemps to avoid projecting our culture on the Greeks. A different picture emerges. The φύσις or physical nature which Thales sought to understand by linking it with water was regarded as a living substance, a kind of magic fluid which contains and imparts life and power. It is, in other words, a mana concept and is said to be a projection of the feeling of the corporate social body. Physis is related to the universe as a human soul is to its body. It is moving, vital, and in man, perhaps in other things, conscious. It is the source in the world of such effects as movement, growth, perception. Thus, the movement of fire or water, growth and activity, thought, all are related to or contain physis. But physis is not

---

2 London, 1912.

merely an uncontrolled daemonic energy. It has a structure; it is presided over by the Moira or Fates who hold it within its limits and organize it into kinds which are manifested, for example, in air, earth, fire, water. The primitive problem is to learn to control this physis in order to satisfy human needs. Magic and religion provided the first techniques. The early philosophers, being the first men who seem to have been aware of the intellectual need, sought to identify it and to use this notion to explain the visible world. Cornford's thesis, which derives from Durkheim and eventually from Kant, is that physis is a projection of the more intense group emotions into primitive man's awareness of the physical world, and Moira is a projection of the group or tribal structure. Thus the physis-Moira complex is a whole to which man, society, and the world, have contributed. It is this which becomes the datum of philosophy.[3] It is by study of this datum that we may come to see how the private world of feelings, sensations, vague dreams becomes the public and structured world of objects within which we lead our lives and about which we may communicate. Of such is the goal of philosophy.

Evidently, though, we are not as yet back to the original analogon of philosophy. Since Cornford's work on this matter, much anthropological investigation has gone forward. The effect of some of this has been to illuminate further the somewhat global notions of physis and Moira as presented thus far.

Some primitive languages consisted generally in a monosound of complex significance, and whose specific kind of reference or function could be changed sufficiently for the occasion by the addition of modifying sounds or inflections. The Nootka language of Vancouver Island is such a language.[4] And, like primitive speech, no doubt primitive thought also consisted in complex unanalyzed concepts applicable with a minimum of change to many situations. Consciousness, concept, emotion, speech, tendency to action, all are merged in one unitary mental experience. Nevertheless, there were some broad distinctions made. Their content, as we may well expect, has to do with elementary mental experiences, the body and its organs, the world and its resistant character in relation to man, the practice of the arts which sustain life: agriculture, hunting, weaving. A vast amount of linguistic evidence for the nature and structure of this thought has been collected

---

3 Ibid., chap. IV. W. Jaeger also emphasizes the continuity of Homeric thought with later Greek thought; cf. *Paideia* (Oxford, 1939) vol. I, p 149 ff.

4 Cf. B. L. Warf, "Languages and Logic", *The Technology Review* XLII, No. 6, (April, 1941).

and organized by Professor R. B. Onians.[5] In the following paragraphs I shall draw, for the evidence for my statements, much from his admirable and careful work.

There is ample indication dating at least from the Homeric age that early man, meditating in his fashion upon the human physis, had come to regard it in some sense as dual. There is no suggestion here of a mind-body dualism; no such distinction was made. Yet a distinction was envisaged between whatever there is in man which is permanent, perhaps immortal—his real self—and the changing, consciously experiencing, waking and sleeping, remembering and forgetting self. Consciousness, it seems, and all the thought, desiring, feeling, which we distinguish within the mind was lumped together and called θύμος. This Thymos or conscious mind was associated with the φρένες which Professor Onians shows was not originally the diaphragm (as Plato thought, *Tim.* 69) but rather the lungs and neighboring organs. The θύμος or thought was identified with the breath that uttered it (the Latin term was *animus*) and with the blood or blood-soul. Even as late as Aristotle, it was believed that conscious thinking was a function of the heart.[6] Perception was, consequently, also a function of θύμος and was associated with the area around the chest. Thus οἶδα, and its aorist εἶδος, ἰδέα referring to perception, sight, and also connoting emotion, a point of interest in connection with the later meanings of these words in Plato's dialogues, are operations of θύμος.[7] Likewise αἴσθησις (sensation) is related to the Homeric αἴσθω, meaning "I gasp, or breathe in."[8] And νόυς, which conceived purposes, was located in the chest or lungs.[9]

The behavior of the θύμος was conceived to be continually changing. It was as manifold as the stream of consciousness, it died out temporarily during sleep and at death when breathing stopped and the heart was stilled it was finally destroyed.

There existed another factor, though, which contrasts strongly with this changing and mortal part of man. The other factor apprehended within the human nature was called the ψυχή or soul (Latin, *anima*, *genius*). This is the part which was permanent. It survived death as the εἴδωλον in Hades and was encountered in dreams, for it was active in sleep. This is the life principle, given

---

5 *The Origin of European Thought about the Body, the Mind, the Soul, the World, Time, and Fate,* Cambridge, 1954.
6 702a, 2-16. cf. Plato, *Phaedo,* 96.
7 Cf. Il. IV, 409.
8 Onians, Ibid., p. 75.
9 Cf. Od. XXIV, 474.

by the gods. This psyche is the water-soul, source of life, power, physical strength, and sex. It is associated primarily with the head, backbone, genitals, and knees. The brain was commonly regarded as the source of the reproductive seed; and other juices of the body, (except the blood), e.g. the spinal fluid, also sweat, tears, fat, marrow, anything wet, were related to the same. Evidently, the importance attributed to the head, even since the Stone-age[10] is thus made comprehensible. Involuntary movements, such as sneezing, blushing, shuddering, as well as the inspirations of the religious genius or poet were attributed to the psyche and were regarded with something of awe as a communication from the inner source of power or from the gods.

There also exists a connection between psyche and δαίμων which is probably important for understanding the religious roots of the drama. Evidently the early Greeks closely identified the two.[11] A daemon, though—e.g. Dionysus,—was always regarded as divine. The belief in the divinity and immortality of the daemon and in the connection of psyche and daemon, Professor Onians suggests, may be at the basis of the Orphic claim for the divinity of the psyche. Plato also connects ψυχή and δαίμων "God gave the sovereign part of the human soul (ψύχης) to be the divinity (δαίμονα) of each one, being that part which as we say dwells at the top of the body . . . and raises us from earth to our kindred who are in heaven."[12] And in the Myth of Er of the *Republic* (617) each individual is assigned a daemon as a kind of guardian angel. In general, then, human psyche and divine daemon are very closely, though vaguely, related. Miss Jane Harrison provides one account of the development of the daemon of ritual into the hero of drama and saga.[13] We shall return to the significance of this connection later.

These beliefs about human nature, as constituted by the semi-divine life-soul or psyche and the conscious but mortal θύμος, or blood-soul, did not remain unchanged. During the fifth century B. C. and after, probably under the influence of Orphic and Pythagorean doctrines,[14] which taught that more of the person survives

10 Onians, Ibid., p. 530-543.
11 Onians, ibid., p. 405, n. 8.
12 *Tim.* 90 A. And at *Tim.* 70A the ruling part of the soul is compared with the sacred image which issued commands from the Temple.
13 *Themis,* Cambridge, 1912, chap. 8.
14 Onians, ibid., p. 115-116. There are, as I have suggested, many recognizable indications of these beliefs preserved in Plato. Socrates, for example, raises the question whether we think with the blood (*Phaedo* 96), and in the *Timaeus* (73C and 81A-B), the brain and marrow are said to be the channel through which contact is made with the gods. But in general, Platonic thought represents a change in these beliefs, cf. F. M. Cornford, *Plato's Cosmology,* (N Y 1952), p. 248-286. The Latins clung to these ancient beliefs more tenaciously than the Greeks. Onians quotes Nonius, "Animus est quo sapimus, anima qua vivimus." (Ibid. 169). Tertullian considers *anima* to be the organ of intuitive klowledge and of Revelation, cf. *De testimonio animae,* 1 and 5.

death than merely the life principle, the notions of psyche and θύμος begin to grow together. The brain becomes something more than a repository for the life seed and tends to become, e.g. with Heraclytus, the seat of consciousness and the personal agent. It comes into a much closer, and perhaps healthier, relation with the conscious θύμος or *animus*, until with Plato, if the myth of Er be taken rather literally, the departed spirits are not merely psyche, bare seeds of life, but they preserve consciousness, memory, power of choice, and awareness of responsibility. At the same time the old fear of the dead and awe of their magic potency loses something of its primitive force.

So much, for the moment, about man and his dual soul. Let us turn to the world, the environment within which human nature developed; how was it conceived? The world is the stage on which man's fate is worked out. What is the nature of fate? It seems to be resasonable to expect, as we have actually found, that the beliefs about the body should be reflected by analogy in beliefs about the mind; similarly men's activity in the world—their arts —may be expected to give the cue to their belief about the world and fate.

This fate, against which the Homeric heroes felt themselves powerless to struggle, is not merely the projected image of tribal organizations nor a personified figure of the boundaries between tribal preserves. The notions which elaborate it go back to an even less complex epoch. They draw their first content, Onians has found, from the weaver's art. Man's fate hangs on "the knees of the gods", for it was on or by the knees that the spinner's spindle was held.[15] The fully developed image is presented in the description of the fatful spinning. Επικλώσαντο θεοί. Fate is made or woven by the gods. In Homer, Zeus primarily, but the other gods also, are those who spin fate.[16] Zeus, however is not bound by fate,[17] but

---

15 Cf. *Republic* X, 616.

16 Onians, Ibid, p. 409-410; 393. Cf. Homer Il. xx, 127; Od. vii, 197.

17 At least in Homer and Hesiod the Moira appear to be subservient to Zeus (Od. iii, 236). **They are the offspring of Zeus and Themis,**—Themis being identified by Jane Harrison as a projection onto the universe of tribal customs respecting boundaries and legal regulations. Aeschylus, however, seems to regard Zeus himself as subject to the Moira (*Prom.* 515). Plato makes them the daughters of Necessity or Ἀνάγκης (*Rep.* X, 617C). Evidently the Moira are agents of a general and pervasive fate whose relationship to other powers, to the will of the gods, to man's free will, is conceived variously at different times.

The root meaning of ἀνάγκη is not clear. Onians connects it with ἄγχειν 'to strangle' which preserves the analogy to a binding cord (or serpent) (Ibid, p. 332), and cites Parmenides, "For mighty ἀνάγκη holds it (reality) in bonds of the δεσμοῖσιν which encloses it around." In the *Timaeus* ἀνάγκη has a kind of pre-rational existence; the chaos which existed before the world was made, has its own mechanical movement (ἐξ ἀνάγκης) which is persuaded to receive such purposive or rational order as it may embody. The same notion occurs in the myth of the *Politicus* (cf. 272D) where Plato pictures fate (here referred to by εἱμαρμένη) as taking over when god withdraws his hand from the tiller of the universe.

he was felt to be morally constrained by his own ordinances and allotments in this respect. Later it is the Moira who weave, and the three sisters just mentioned may be personifications of parts of the weaving process. Lachesis selected the wool, (weighed out for her by Zeus himself), Clotho spun the thread, Atropos wove the web. The fate thus woven at or before a man's birth is then thought to bind him as if it were literally a web or a cord. In this manner fate controls man's life.

A man's fate is bound upon him by πείρατα. The term which becomes so important in Plato[18] was used in connection with spinning. It means thread or woof-thread, a spun rope, or even a knot. It came also to refer to the cord or thread by which a man's fate was fastened upon him and whose presence is betrayed by its effects.[19] No doubt a cord or thong with which a man could bind things together, a bundle of sticks, a group of prisoners, so that he could treat the many as one was a most valuable instrument and an impressive source of power. To the primitive mind it could well seem to be a god-like instrument suggestive of the way in which the universe is held together. Later πείραρ came to mean boundary, limit, or form, approaching the sense in which Plato used the term. The figure of fate as a bond or ligature which limits or circumscribes a man is expressed in many ways. One recalls the ring binding a marriage, a king's magic ring, the collar signifying slavery, a belt with its many meanings, and especially the crown of oak or ivy leaves or a kingly crown which quite literally binds the psyche.

An image which runs through Homer suggests that in the web which binds a man, the warp threads are time or length of life. This time is to be understood as qualitative time, time as lived, its hours differentiated by the varying contents of experience. The woof threads are fate, the ordained events themselves. The tapestry thus woven and bound upon a man, by design of Zeus, is unchangeable. Yet within limits, undefined limits it is true, there was sometimes conceived to be freedom. A man could render his fortune worse by his ill choice, but just how and to what degree choice could affect fate was a problem which ancient man could scarcely reckon with.

Much of this lore concerning man's relation with fate is summarized in a cosmic figure of some interest. Okeanus, over which

---

18 E.g. *Phil.* 23f. In *Politicus* (309B-C), the kingly artist is said to use true opinions as the bond with which he binds the soul to the divine. (The term used here for bond, τὸ ξυγγένες, is also used in connection with spinning.)

19 Onians cites the Greek epigram, "Such is the wretched life of mortals, so unfulfilled their hopes, over which the threads of the Moira hang." *Ibid.*, p. 337.

the heavens bent like the inside of an egg shell, was the primal sea from which,' as Thales recalled, the world was born. It was said to lie around earth's shores like a serpent.[20] As the source of the world and its life, this water, Okeanos, is the psyche of the universe and the source of the human psyche. At the same time, since it lies around the world, limiting it and holding it together, it is the world's fate. Thus at this point the notions of life, innermost nature, and fate become conjoined in a manner which presages later beliefs.

The convictions of the primitive Greeks which I have just described are shown by Onians to have been extraordinarily widespread. Like primitive tools, primitive beliefs seem to have been all of one stamp the world over. This is an astounding fact. Why this is so we can only speculate.[21] Perhaps once a single culture came to dominate a people who later spread and populated or conquered the world. Perhaps all peoples faced with a given type of problem at a given stage in evolution react in similar ways. Whatever the explanation of this fact may be, the prevalance of this body of beliefs is ancient enough and is sufficiently widespread and its content sufficiently explicit to justify our suspecting it to be the archetypal datum of men's thought.

My point to date is that Cornford's findings and his generalizations on them are supported and considerably enlarged by Onians' investigations. The result of these studies is that an image emerges which quite evidently dominated and directed ancient speech, belief, and activity, and which is by no means without later and present power, although it has been variously interpreted and thereby transformed. This is the image of a man exercising a simple art and at the same time, through its agency, coming in many other ways into significant contact with the world. The art is spinning and weaving cloth of wool. But from this simple practice a universe is born. The cloth, as it magically takes form under his hands, not only has values which preserve life, but it acquires meanings which relate him and it to the animal life from which its material came and to the environing universe itself. The cloth is not only woven but interpreted. It becomes the web of a mysterious fate within

---

20 The serpent, as every one knows, is a phallic symbol. Also it is a symbol of death. Snakes were supposed to haunt graves and perhaps to be a form which the dead assumed. Onians notes the ancient belief that the spinal marrow, repository of psyche, was thought to turn into a snake after death, a belief deriving from the resemblance between the spinal cord and a snake's skeleton. Ibid., p. 206. Socrates twice refers to Homer's mention of Okeanus as the origin of the gods and connects the legend with Heraclytus' philosophy. Cf. *Crat.* 402B; and *Theat.* 152E.

21 Cf. J. Campbell, *The Hero with a Thousand Faces*, N. Y., 1949, on the common character of myths.

which he himself and his humble weaving are woven and ordained images and within which he is to make his place and his peace. Likewise, as the web is at once a manifest and useful object in his hands and also expressive of the unmanifest and awful fate beyond, yet including the factual present, so he too, the weaver, is not merely a conscious active and thinking self (a θύμος) but is also a something or a part of something (the ψυχή) which touches on the same mysterious source of life and power which he must somehow use and control. Against this background of belief, Plato's image in the *Politicus*[22] of the statesman weaving the two different kinds of human souls into the fabric of the state, takes on a rather special interest and grandeur and opens the way, perhaps, to rendering the ancient fate somewhat less awful and less blankly mysterious.

We have insensibly moved in this discussion to the point where interpretation rather than the content to be interpreted is the point of emphasis. The content, in sum, is a dual-natured man involved in the toils of fate and dependent upon its source. What, we may now proceed to ask more explicitly, was the interpretation put upon it during the epoch when it was accepted as common sense?

For the primitive man the practical was undoubtedly prior. Hence, one would be disposed to expect that the kind of interpretation which in ancient times was placed upon this subject matter would be directed toward enabling man, as factored into the two souls, to accommodate himself efficiently to the flow of events as factored into fate or a compelling order of some kind and the gods or the source of that order. And it seems undoubtedly to be true that the primitive man is not concerned with his own mind or with the cosmic weavings for their own sakes. He is by no means the natively curious, experimentally minded and poetic man which some romantic biographers have imagined. Undoubtedly tradition and custom bound his every thought and movement. His real problem is the task of finding the strength and supplies needed to remain alive and to insure the welfare of tribe and clan and in addition to dealing in some effective fashion with the anxieties which the many frustrations of his hard life bred in him. As we might express it, he is more concerned with a method for reckoning with his fate than with understanding it. His interpretation of his life is rooted directly in this practical problem. Also his method

---

22 279B and passim. Also compare the remark in the *Sophist* (259E), "any discourse we can have owes its existence to the weaving together of forms". (Trans. Cornford)

of interpretation is not something which he is to invent or is at liberty to change. It too was something laid down from time immemorial. His is the religious method. Religion provided the interpretative context within which his picture of man and the world was rendered functional and effective. My point, to revert back to the thesis of this paper, is that this religious interpretation added elements to the content as developed thus far which make the whole complex recognizably the datum of philosophy. Let us turn, therefore, to consider the relation between rite and myth and the beliefs just outlined. We shall see that myth and rite contain—by inference, of course—a theory of the functional connection between man and fate which is sufficiently fertile to lend itself later to philosophic use.

It is quite difficult to generalize about the kind of relation which formerly was supposed to hold between man and fate. Did the primitive man believe nature or fate to be such that he could forceably control it by his magic, myth, and ceremony? Did he think to appease or to bargain with the overpowering forces around him? Or did he expect by these means rather to bring himself into harmony with fate? We may say that the techniques of his religion—ritual and myth—were the means by which primitive man actually adjusted his conscious self and activity to unconscious drives within him and to the forces of the external world. Clearly, though, primitive man made no such explicit theories. He expected that if he did and said exactly what had always been done and said that his welfare would be assured. In particular, it is not probable, contrary to what has so often been suggested, that myth should be regarded as primitive science. Lord Ragland has argued, quite conclusively, in my opinion, that myth is neither legendary history, pseudo-science, nor an original product of the ancient poet's imagination; it is the thing said as accompanying or as part of a ritualistic action,[23] and translates into the linguistic medium the character of ritual. The myth might be said to be the formula or description, and finally perhaps the explanation, of the rite. This is an important point, for the myth is communicable and has survived. Through interpreting it aright, therefore, we may reach some insight into the primitive mind and its springs of action which can throw valuable light upon that which has been called the archetypal datum of philosophy.

The intent of the myth-ritual is always to produce what is necessary for the continued life of tht tribe, protection, food, cloth-

---

23 *The Hero*, N. Y. (Vintage), 1956, part II.

16   TULANE STUDIES IN PHILOSOPHY

ing, sense of security. It is, therefore, regarded as the source of
the tribe's life and well-being. Attention has already been called
to the remarkable similarity in form among primitive myth-rituals.
All project, as some term it, the life and identity of the tribe into
a king-figure who then becomes the central personage in the ritual
and the key to the tribe's well-being. Though he may be called
by various names, the pattern through which his life is communicated is always recognizably one and the same pattern.

A schematic account of the early ritual worship of Dionysus
consisted in the following movements: [24] (1) an orgiastic procession
into the country-side area associated with past occasions of the coming (or birth) of the god; (2) the dismemberment and sacrificial eating of the god or of his appropriate totemistic surrogate; (3) the triumphal procession of the rejuvinated and enthusiastic worshipers
back again to the city where, likely enough, a marriage ceremony
of the resurrected god and some favored maiden would be enacted.[25] These divisions correspond with the kinds of dances which
continued to be associated with the Greek drama, the chorus moving through their patterns on the konistra. These dances are,
(1) the circular-form dance of protection, danced in the early
ritual by the Kourites around the madonna and new born god;[26]
(2) the dance of lamentation, bemoaning the death and dismemberment of the god; (3) the fertility dance associated with the
resurrection and the marriage of the restored god.

The details of the lives of several ritual kings,[27] Osiris, Oedipus,
Theseus, Dionysus, Zeus, Moses, Siegfried, Arthur are set forth
by Lord Ragland and compared.[28] They are shown to be similar
in almost all important points. These important points fall into
three classes: those connected with the strange and obscure birth
of the hero; those which relate to his accession to power, the won-

------

24 Cf. Euripides, *The Bacchae*.

25 S. H. Hooke provides a somewhat different summary of the general pattern.
He writes, "This pattern consisted of a dramatic ritual representing the death and
resurrection of the king, who was also the god, performed by priests and members of
the royal family. It comprised a sacred combat in which was enacted the victory
of the god over his enemies, a triumphal procession in which the neighboring gods
took part, an enthronement, a ceremony by which the destinies of the state for the
coming year were determined, and a sacred marriage.

"Together with the ritual and as an essential part of it there was always found,
in some form or other, the recitation of the story whose outlines were entacted
in the ritual. This was the myth, and its repetition had equal potency with the
performance of the ritual. In the beginning the thing said and the thing done were
inseparably united." S. H. Hooke, *The Labyrinth*, London, N. Y., 1935, p. v.

26 J. Harrison, *Themis*, ibid, Chaps. I, II.

27 Cf. J. G. Frazers' account of the Rex Nemorensis, *Golden Bough*, (abridged
ed.) N. Y., 1942, Chap. I.

28 Ibid, Chap. xvi. Cf. A Toynbee, *A Study of History*, Oxford, 1939, vol. 6,
p. 476 f.

ders which, amidst difficulties and threats, he works for man, and his violent death; and finally those which tell of the holy place of his burial and his resurrection or continued potency. This pattern is closely similar to that which Van Gennep has found to be present in the large group of ceremonies which he terms rites of passage.[29] These are the rites which commonly preceded any change in life which called forth a need for new strength; e.g. birth, initiation, marriage, war, sickness, consultation of an oracle, death. The god of the ritual, the hero of the myth, is preeminently connected with the passage of the year from winter to summer and fruitfulness, and similarly with the tribe's passage from danger to safety, from threat of starvation to plenty, with the crops' winter death and summer harvest, with the animal's (and man's) passage from sterility to fruitfulness. In short, he is connected with—or rather *is*—the movement from one form of life, through danger and death, to renewed life. This movement, evidently, is the ordinance of fate. Fate, thus, moves in cycles. Ancient man met the fact of death and danger by interpreting them as a crisis to which his fate had bound him but through which it might also carry him by the religious means which was, in effect, his own attempt to imitate fate.[30] This pattern of ritual is evidently expressive of primitive man's sense of empathy with the universe.

This ritual-myth can now be related back again to the beliefs which the study of such words as ψυχή, θύμος, and ἀνάγκη rendered explicit, and it will be seen more clearly that ritual interprets these beliefs. Two useful clues to the nature of this relationship exist. One is provided by the Greek drama, the other by late survivals of the mystery religions. If the connection between ancient action or ceremony and beliefs about man and fate can be made clear, it should not be difficult then to show how they developed by continued re-interpretation into philosophy.

It will be appropriate first to consider an account of a survival of the mystery religions. I have in mind Pausanius' description of the rites of Trophonius at Lebadeia. Although this account is relatively late in date (c. 147 A.D.), there is good reason to believe that the rite itself is very old indeed.

---

29 *Les rites de passage, Paris,* 1909. Cf. also *Aeschylus and Athens,* George Thompson, London, 1941, Chap. VII, also p. 380.

30 Thompson (op cit chap. 7) thinks that the purpose of ritual is not so much to harmonize a man or the tribe to fate as it is actually to break the old bonds (πείρατα) and to get an entirely new set of them. In the case of the individual, this fate was thought to be changed by change of identity, name, etc. Thus rebirth was a means to getting a different fate. Whether this interpretation or the one developed above is correct is difficult to say, nor does it matter much for present purposes. I am willing to conclude that ritual was the means either for learning how to cooperate more efficiently with fate or for changing it.

Trophonius was a river god, a god of fertility. The serpent was one of his symbols; goats were sacrificed to him. The cave at Lebadeia, which was his sacred place, was also the source of the river Hercyna. Many others besides Pausanius, came here to consult his oracle. Pausanius tells[31] of the preliminary rituals,— his bathing in the river Hercyna, his eating of meat from the sacrifices, his sacrificing a ram, then drinking of the water of Lethe and Memosyne (a preliminary cleansing of the soul), his worshiping the god's image, and finally his entry into the Oracle itself. The Oracle is a cavern shaped like a bread-oven with a chasm within it and a hole in the floor of the chasm. The inquirer enters the hole or mouth and thence is caught up as if by an eddying current. Pausanius continues, "After this those who have entered the shrine learn the future, not in one and the same way in all cases, but by sight sometimes and at other times by hearing. The return upwards is by the same mouth, the feet darting out first."[32] The priests first question him concerning what he has learned . Then "after gaining this information they entrust him to his relatives. These lift him parlyzed with terror and unconscious both of himself and of his surroundings and carry him to the building where he lodged before with Good Fortune and Good Spirit ($\check{\alpha}\gamma\alpha\theta\circ\varsigma$ $\delta\alpha\acute{\iota}\mu\omega\nu$). Afterwards he will recover his faculaties . . . What I write is not hearsay; I have myself inquired of Trophonius and have seen other inquirers."

It seems inescapably evident, from this account, that the process of inquiring of this Oracle—as no doubt of the others—follows the ritualistic patterns described above. There is the period of purification which separates the inquirer from the things of the world (compare the strange birth of the mythic hero) and prepares him for his rather special role. Then there is the actual struggle at the threshold and the encounter with a supernatural power inside the oracle whence comes the new knowledge, analogous to the god-hero's struggle to manifest his power. And finally, there is the return to the ordinary world plus, though, the new accession of insight under conditions which strongly suggest a kind of rebirth. This is clearly the account of a mimetic death and rebirth with new powers. It is not too much to say, perhaps, that we have here an indication of the ritual method by which ancient man, and man not so ancient too, sought to regain contact with the sources of his life and to acquire therefrom the forces and knowledge which

---

31 *Description of Greece*, ix xxxix Loeb vol. IV, p. 347 ff. Cf. also Plutarch, *Moralia*, xxii;—
32 Ibid. p. 353.

he needed to help him move through the changes of life. He sought to use death as a means for coming into a more lasting touch with life. Reverting to our former term, I would interpret the contact supposedly made by Pausanius, to be a contact between his psyche and the god, for the psyche is that part of him which traditionally could communicate with such a daemon as Trophonius. This contact was expected to touch on his daily life —by mediation of the thymos—and to revivify it in such fashion that he could either deal more fruitfully with fate or find that his fate, like the new year, was entirely renewed. In such a manner as this, the ritual served to interpret beliefs about man and fate by bringing them into effective touch with life.

The tragic dramatists do not radically alter this theme. That this drama is derived from the ritual which we have been discussing has been ably argued.[33] Since the ritual did develop into the tragic and comic drama, the nature of this drama will provide a trustworthy indication of the character of those elements implicit in the parent ritual which were useful enough to survive. The most important of these elements is undoubtedly the tragic pattern.

It has appeared to me that the essential elements of the rather complex pattern which the drama follows can be easily and briefly summarized, as Aristotle suggested, in terms of the beginning, middle and end of a drama.[34] The beginning is the hero's identification of himself as a person able and resolved to play a role which he believes to be appropriate. The middle recounts his attempt actually to play this role and his finding that these attempts meet increasing resistance, for he finds himself engaged in a struggle— the agon—with what he takes to be malicious and destructive forces. The end comes when this resistance is recognized to be fate, or something like it, and the hero is brought to recognize in consequence of this opposition from fate that he has misidentified himself and his role in life. The end clearly is catastrophic, and at the same time it is a mimetic rebirth. For the insight to which this crisis has led reveals the illusory nature of the identity which the hero had first assumed for himself and enables him, therefore, to

---

33 Cf. F. M. Cornford, *The Origins of Attic Comedy*, Cambridge, 1934; Sir Gilbert Murray, "An Excursus on the ritual forms preserved in Greek Tragedy," in *Themis*, by J. Harrison, Oxford, 1912, Lord Ragland, Ibid, part III. Cf. S. H. Hooke, Ibid, p 31f where the tragic chorus is connected with the protective dance reproducing the pattern of the labyrinth which hid and protected the community's life-renewing king or totem god. G. Thompson, Ibid, p. 382, summarizes the connection between the god of the ritual and fourth century actor thus: "They (the actors) were mediums for expressing what had once been the voice of a god. The actor who spoke the part composed for him by the poet was descended from the poet-actor; and the poet-actor, who spoke the words which he had been inspired to compose, was descended through the leader of the dithyramb from the priest at the head of the thiasos, who, since the god had entered into his body, *was* the god."

34 Cf. my *Art and Analysis*, (Nijhoff 1957), Chaps. 9, 10.

reach a more harmonious adjustment between himself and fate. This pattern fits almost any of the tragedies so obviously that an illustration is scarcely needed. Thus a tragedy is the presentation of a movement through suffering to wisdom.

It is obvious, though, that the ritual pattern has undergone a profound and civilizing change. Miss Jeanne Duchemin[35] finds reason to believe that the tragedians took their cue from the argumentation which took place in the courts of law in Athens. Here in effect was an agon, but it was a struggle in which words and reasonings replaced the human sacrifice and bloody dismemberments of bygone milenia. Persuasion and understanding thus came to replace force and magic, and they achieved a reconciliation hardly possible to the latter. It was A. N. Whitehead who noted that civilization began with the intuition that persuasion, not force, is the divine agency in the world.

In the *Orestia* Aeschylus celebrates the *rite de passage* from Clan morality or the vendetta, to the civil morality of the Greek city-state. His emphasis seems to be placed upon the social effects, the group-persuasiveness, of this movement through crisis to rejuvenation, which again is the story in brief of the substitution of insight and rational persuasion for force and blind tradition in the ordering of groups. In Sophocles the emphasis is placed upon the importance to the individual of the movement through this pattern and the insight achieved, the reformation of character, and the approach to peace which follow therefrom.[36]

The tragic dramatists have taken a pattern common to the death-resurrection ritual associated with the mystery religions, the primitive goat-songs, and with the rites of passage, and have converted it into a means whereby a society or an individual may move through decision, action and suffering, to a rejuvenating insight into himself whereby he may adjust himself more harmoniously in thought and action to fate. It is as if the Hero maneuvered himself into a situation in which, through the workings of fate, he was able to achieve a clearer insight into his less clearly known self (his $\psi v\chi\acute{\eta}$) and then through this wisdom to become better able consciously to cooperate with his own fate. The tragic hero's

35 The ΑΓΩΝ *dans la Tragédie Greque,* Paris, 1945, Cf. Carl Conrad *the Tragic Conflict in Aeschylus,* unpublished thesis, Tulane University, New Orleans, La., 1956.
36 The *Eumenides* presents a dramatization of the change in the idea of fate from the older notion of an inviolable even mechanical order, in which the Homeric man was entangled, to that of a moral order in which not the act alone but also the circumstance and the motive were significant. The Errinyes themselves evidently, are the protagonist of this drama. Sophocles' *Antigone* dramatizes the opposition which sometimes relates $\psi v\chi\acute{\eta}$ and $\theta\acute{v}\mu o\varsigma$. Antigone represents the intuitive mind which seems to come into contact with a truth and a law which is beyond the human; Creon's is the logical mind meeting problems with the techniques of this world. Together the two plays offer a beautiful presentation of the complex notions of fate and mind and their interrelation through action and ritual which constitute the philosophic datum.

noetic problem is to know with conscious clarity both himself and the world in which he has to act. His ethical problem is to express this knowledge in his behavior so as to achieve a harmony between himself as he actually is (his Psyche) and the exigencies of the real world (ἀνάγκη). There is more than a suggestion in the tragedies that this is a problem which no one ever solves wholly and finally. Any solution is partial and ends by producing a new problem. Such is the irony of the wisdom got through human suffering. Oedipus of *Oedipus Rex* learned at least who he was not, but pride at this achievement, together with a certain degree of self-pity, led him to Colonneus . . . When the human psyche was renewed by such insights as those which the Greek drama has to communicate, then philosophy was ready to begin in the world.

The tragedies, then, present a mature and artistic use of an ancient ritual form and its content of mythic belief. It seems inevitable that Plato should have been deeply affected and, whatere his strictures upon their effect upon audiences, deeply influenced by them. We know, for instance, that Plato was fond of the little plays of Sophron. Also he expressly compares a section of the *Politicus* (cf. 303C) to a drama in which the statesman-hero has got himself lost in the satyric thiasos and must be separated therefrom and identified by the dialectical means. And he remarks in Laws VII (817A) to the tragedians who visit his state, "we also according to our ability are tragic poets, and our tragedy is an imitation of the best and noblest life, which we affirm to be indeed the very truth of tragedy." Evidently his rejection of the drama is infected with more than a little irony, for it is evident that the tragic ritual underwent another transmutation in his hands, such was its power, and emerged in the highly intellectualized form of the Platonic dialogue.

One of the Platonic dialogues which illustrates as well as any how more complex and communicable interpretations are built out of mythic material is the *Phaedo*. Here the myths of two cities, Athens and Thebes, are subtly woven together. The Thebans, Simmias and Cebes, have journed from the city of Cadmus and Harmonia to hear the condemned old man from the city of Theseus elaborate and expound their own Pythagorean doctrines.[37] The rites connected with Theseus' ritual slaying of the old king in bull form within his labyrinth have already caused the postponement

---

37 Socrates is represented as extending the theory of forms explicitly to include the virtues (*Phaedo* 74D, 83). But of course, the generalized question about the extent of the forms does not arise until *Parmenides* 130. Some scholars hold that this ideal theory is elaborated in this dialogue for the first time, being only adumbrated in the earlier ones (cf. F. H. Cornford, *Plato and Parmenides*, (N. Y., 1951), p 74; 83 n.

of Socrates' death.[38]   Socrates refers to his composing a hymn in
honor of the god (*Phaedo*, 61A).   Theseus, once King of Athens,
whom the Athenians imagined to be miraculously present at the
battle of Marathon, is one of those culture heroes, chthonic daemons,
who brought life and the arts to men.   One of his more famous
exploits which, as Plutarch says,[39] everyone know about, was the
expedition to Crete where he led the fourteen youths, with the
aid of Ariadne's thread, safely through the labyrinth, killed the
Minotaur, married the queen and preserved the life force and
celebrated his success in a dance imitative of the windings of the
labyrinth.[40]   One is almost tempted to read the legend as an allegory.
Socrates is a Theseus who leads the fourteen participants of the
dialogue (*Phaedo* 58A-59A) through the labyrinthine dialectic,
following the golden thread of the ideal doctrine, to slay the Minotaur which may be misology (ibid 89D), or may be the hobgoblin
death (ibid 77E), or it may even be the Great Beast of the *Republic*.
Finally the myth of fate which closes the dialogue presents again,
with its intertesselated rivers and routes, a kind of cosmic labyrinth through which only the lover of wisdom is able to find the
way.   Such allegorizing, though, is rather too easy.   Besides it
leaves the Thebans to one side, and their function is essential.
Indeed without their questioning of Socrates, there would have
been no dialogue at all.

Simmias and Cebes are interested both in the ethical use
which Socrates has made of their Pythagorean doctrines and in
observing the effect which these doctrines have had upon his
character.   They have, therefore, come to try Socrates.   However,
Socrates' questioning not only destroys the arguments of Harmonia
and Cadmus (the legendary founders of Thebes, ibid, 95), but
suggests that Simmias and Cebes have not sufficiently used and
examined their own beliefs.   One is led to suspect that Simmias
and Cebes have specialized in the mathematical part of the Pythagorean doctrine (which partly accounts for Socrates' many mathematical observations and analogies) and have neglected its more
philosophical and ethical parts.   In consequence of their specialization, they have not conceived the nature of the soul in harmony
with their own doctrine.   Their conviction concerning the eternal
character of the mathematical objects which they understand (and
hence have in their souls) is in conflict with their fear of death
and their conviction of their own mortality.   In fact, as A. E. Taylor

points out,[41] their doctrines recall the epiphenomenalism and interactionism of a later mathematicizing philosophy. No doubt any Pythagorean would be hard pressed to explain how a soul which could be accounted for by either of these two doctrines could do mathematics, much less achieve immortality.

Evidently, specialization in mathematics is not sufficient for salvation, for the consequence of specialization is an unjust neglect of that part of the soul which the specialist fails to cultivate. Socrates' own doctrine seeks to avoid this injustice by uniting the pursuit of virtue and the good represented by Athens and the Theseus myth with the pursuit of mathematics represented by the Thebans; the soul is constituted by its eros for the good, which is its life, as well as by its recollection and vision of the mathematical forms which, rightly used, are a means to the good of the whole. Specialization in either of these to the neglect of the other is a kind of Sophistry and the way of death, for one cannot treat or cultivate the part in isolation from the whole of the soul (cf. *Charm.* 156E) nor the soul in isolation from that larger environing nature which contains and nourishes it.

Thus the mythic and dramatic overtones of the dialogue are integrally a part of it and present in a concrete manner a story of the nature of the soul and its fate which the remainder of the discussion interprets more abstractly. The dramatic development converges clearly upon the point that Socrates, the contemporary Theseus, is an embodiment of Philosophy, and that his explicit doctrine is an interpretation of such a life and that the crucial point of this doctrine is the inclusion of both the Socratic vision of "the containing power of the good" together wth more precise and communicable mathematical and naturalistic speculations into one harmonious whole. Thus the dialogue is a transformation of the myth-ritual, and by its express inclusion of the myth along with the more abstract doctrinal interpretation, it provides an instance of the philosophical practice which it recommends. Plato, I suggest, would agree whole-heartedly in Jaeger's remark: "Mythical thought without the formative logos is blind, and logical theorizing without the living mythical thought is empty."[42]

Judging from this sample suggestion and the many others which will occur to anyone familiar with the dialogues, it becomes quite evident that Plato owed a vast debt to the mystery ritual and to the drama.[43] It is quite clear, in general, that in the earlier

41 Plato, *The Man and His Work*, (N. Y., 1956) Meredian Books, p. 198f.
42 *Paideia*, (Oxford, 1939), Vol. I, p. 150.
43 Cf. Harrison, Ibid, p. 513. Also Er's journey in *Republic* X, and his account of the fate of every soul. The whole of the *Republic* submits without forcing to being read in relation to the myth which is mentioned in book I and to the myth of Er.

dialogues a cathartic dialectic provides the torpedo shock which helps to free or purify the inquirer from convention, and then to elicit anamnesis through which he comes into relationship with the eternal forms; in the end this insight is brought again into the cave of the world and used to illuminate daily activity and obligations. One might say, in these terms, that the practical problem for Plato is the problem of disciplining and purifying the mind so that it can learn to use that part of human nature which can be brought into touch with the forms and then to communicate the insight thus gained for the end of controlling and enriching life and especially for achieving a still better insight into that life.

If this is so, then we sould expect one of the characters in a Platonic dialogue to move through the tragic (or comic) pattern by accepting a role—perhaps as a believer in a conventional definition of virtue or knowledge—then in the struggle with Socrates or others, who mediate or represent fate,[44] to find that role contradictory and impossible, then finally to achieve a positive insight into himself, especially into the ruling part of himself, or at least to accept the negative knowledge and recognize his own hybris and actual ignorance. Thus Socrates points out that the function of the philosophic dialogue is to enable the participants "to escape from their former selves and become different men" (*Theat.* 168A. Trans. Conrford). Such, again, is the cyclical movement through which the conscious man comes into contact with his enduring self and with fate beyond himself or with the source of fate. No doubt Plato's immediate philosophic problem is to come to understand as clearly as possible the nature of this enduring self and the cyclical procession of the gods within which it has its being, and further, remembering the tragic character of all human effort, to understand so far as possible his own failure to understand. Perhaps in this manner he converts tragedy into comedy . . . .

I conclude now with a summary and a few observations. This paper is intended to provide a likely story concerning the philosophic datum. This datum is archaic experience, the earliest complete (from our point of view) interpretation of the human situation. The conviction that man's nature is dual is extra-ordinarily ancient and persistent. His conscious self was thought to be different from and variously related to a more primitive, powerful, and enduring self. An equally well-tried conviction held that men are situated within a world-order or fate which extended beyond themselves; to this was usually added belief in a source of this fate, the gods,

---

44 Socrates is said to be a spinner of fate in *Theat.* 169C.

upon which both it and men were dependent. Thus man and his fate provided the content and the problem for antique thought. The early interpretation of this content maintained that the self of ordinary experience depended for its well-being upon the world around it, and this dependency was a function of the kind of relationship which it established with its psyche and, through this, with the gods. The gods ordained fate, and at the same time they conveyed to men's psyches the intimations of their own fates. These intimations could be translated into action and attitudes which were acceptable to the gods, and hence were harmonious with fate, by means of the conscious mind, the thymos. Thus the way by which one came into contact with the psyche and with the powers beyond fate was ritualistic. Ritual established a continuity among the several otherwise discrete elements which entered into this complex.

I have also tried to indicate how the Greek drama proceeded to develop the means of interpretation, that is the ritual and tragic pattern, as well as its content to the point where these become an instrument designed as if by nature to the hand of Plato. I believe it to be not too difficult to show that succeeding philosophic history elaborates, clarifies, and sometimes simplifies or omits parts of this initial datum. As it seems to me, there are perhaps only three crucial points in the history of the use of this datum. The first occurs, as I briefly indicated, when Plato intellectualized the ritual pattern, converting it into a means for coming into contact not with the gods and their mana but with the intelligible forms. Other elements of the datum were, of course, altered accordingly. The crucial stage is represented in Descartes' philosophy as well as in any, for here the exclusion of moral and value aspects from fate and its re-definition as a mechanically determined order becomes quite clear and persuasive and the problems which follow thereupon not less clear and insistent. And finally in Kant's critiques the interdependence of the knower and the known, of the self and the world, sets the sage for modern developments.

There is, then, a complicated, changing, and perhaps developing set of relations among ψυχή, θύμος, ἀνάγκη and the gods or the source of fate. Interpreting the nature of this complex and the character of the actual and desirable relations among these elements of the datum is evidently identical with the search for the thread which will lead through the labyrinth of history. And further, the task is evidently to be accomplished by cultivating those crises which lead through dialectical struggle to the requisite insights.

Perhaps this is merely to say that a man must integrate himself and adjust to his environment. I think, though, that the present approach says something more. In particular, it suggests the depths and complexities in human nature and the human situation which seem to be missing in the platitude. It also suggests a way of designating the philsophical datum which may avoid certain hazards. For if the material of philosophy is to be recognized in myth and if myth is regarded as a likely story having indefinitely many interpretations, then the philosopher is apt to accept the obligation of remaining in touch with his origins and to preserve a healthful attitude of Socratic ignorance in his dealings with this oracular material. Such an attitude, no doubt, most philosophers cultivate and desire to preserve no matter what function their philosophizing performs, whether analytic, critical, or constructive. The consequences of this relation should be beneficial in more than one way. For one thing philosophy would be likely to accept its relation to the whole of its originating datum and hence to attempt to do justice both to man in his complexity as well as to the nature or fate which surrounds him. Consquently, philosophy would not center its interest solely upon the human being nor solely upon the world and its sciences but would seek the whole of its destiny through apprehending both of these and their interrelations. For another thing, philosophy viewed thus, remembering the unlimited fertility of its mythical matrix, would not risk arresting the process and powers of inquiry and interpretation by erecting complete systems. It would probably look upon the supposedly all-inclusive systems which some thinkers have attempted to construct as so many stories in the Tower of Babel, useful primarily so far as they have provoked criticism and fruitful rebellion. It would recall, too, that philosophy as taught in the schools and as written in books is merely a meautic technique for bringing philosophy to birth within men, and that erudite terminologies, learned "isms" and all philosophical doctrines are never more than ironical means to this end.

Evidently the primary obligation of the philosopher is to respect his subject-matter. It is not to take sides in contemporary controversy and defeat his opponent, nor to construct an elenchus-proof system within which he may take refuge. Rather he expresses respect for this subject matter and enters effectively into the philosophic *agon* by keeping open the ways of interpretation and philosophic conversation and by this means continually exploring and illuminating the sources of conflict and resolution, of blindness and insight.

PHILOSOPHIC DISAGREEMENT AND THE STUDY
OF PHILOSOPHY

Richard L. Barber

FOR all those who regard the history of philosophy with serious respect, two facts seem inescapable. The first and more obvious one is the multiplicity of philosophical systems; the second, —less evident, perhaps, but I believe equally unassailable,—is the moral earnestness and intellectual competence of the thinkers who have devised and bequeathed the great majority of these systems.

I

I propose in this essay briefly to examine these two "facts", their evidence, and some arguments about them. Then, if my considerations to that juncture seem appropriate, I shall ask you who read this to examine with me some implications regarding the nature of the philosophic enterprise, in fact and in principle, which, I believe, follow from these two aspects of its history. Finally, I will suggest in outline what appears to me an appropriate attitude for the student or teacher of philosophy toward his subject-matter.

II

There is little need to call the attention of even a beginner in the study of philosophy to its plurality and diversity. Almost before he learns, and well before he understands, what they have said, he learns that "philosophers disagree". With each other, with himself, with thinkers in "other" disciplines, with their own tentatively advanced speculations,—they disagree to an extent and with a depth of probing unencountered elsewhere. Indeed the beginner in philosophy may be disenchanted on this very account, convinced that out of such variegation can emerge no constructive truth or even advance toward truth. As he progresses in his early studies this conviction and its ground in his experience are likely to grow; he may swing between occasional extremes of exultant dogmatism in some new truth and despairing abandonment of the whole enterprise; but his more likely and frequent responses will be the growing awareness of the variety of this activity called philosophic, and the suspicion that it may all come to no good end.

This situation is sometimes terminal, unfortunately. There are some escapes from it that are no less unfortunate, in my opinion; and there seems to be at least one good way out.

The "unfortunate" exits include the following: (1) The conclusion that all the divergent philosophies are false and wrong but one, that one being right and true. (2) The rejection of all philosophy as vain and fruitless. (These are the two "occasional extremes" just mentioned, but now crystallized and elected terminally; they are much less subtle and worthy even than the alternatives which follow.) (3) The decision that all the seemingly diverse philosophies are not really so opposed, but are *somehow* in agreement in their concerns and conclusions. (4) The confident declaration that this agreement has at last been made clear by the researches of the declarer, whose own system now triumphantly accepts, subsumes, and displays as facets the work of his historic predecessors (and also, by generous invitation, makes room for the fragmentary insights of lucky successors who may happen on something he has overlooked).

What makes these escapes unfortunate, in my opinion, is that they all involve the repudiation of the *history* of philosophy. The holder of any one of these views has ceased to regard that history and its contents with the "serious respect" which was asked in the opening sentence of this essay. For this it seems necessary to accept philosophers and philosophies at their own appraisals, to hold at least tentatively that they are what they *say* they are, and do what they *say* they do; and quite obviously none of the four alternatives just described can so accept them, as instead these would haughtily reject or happily subsume that which never meant itself to be rejected or subsumed.

The better way out, and about the only one left, is to accept the diversities at face value and learn to live with them. That this will require a careful statement regarding the nature of philosophy

I do not deny; and if this statement but add to the divergence and disagreement, I can not by my own principles demur. I can only hope that it will offer something in interest or insight to you who read it.

III

The second consequence or concomitant of a serious respect for the history of philosophy was asserted to be an awareness of the integrity of the thinkers responsible for the content of that history. To explain and defend his claim it will now be necessary to say what, in my opinion, these philosophers thought they were doing. It seems clear to me that all systems of philosophy have two common aspects, if we consider them abstractly enough; first, the subject matter or area of concern for each is the "all-of-reality-that-can-be-known", whatever or however much that may be according to that particular view; and secondly, it is the aim or intent of each philosopher to express truths of the utmost importance and scope regarding this "all-of-knowable-reality"; if my language should seem extraordinarily ordinary here, it is because I am trying to speak no man's jargon, lest I be myself 'subsumed' in the waiting net of his system.

These two aspects then are asserted to be discernible in every philosophic system: the totality of the interest, and the ultimacy of the intention. Further, I would think that every one of the two or three hundred thinkers and writers whose work is currently deemed deserving of our sustained study was quite able to think of his enterprise in these terms, and did so think; most of the systematic writings contain adequate evidence of such awareness, although most often expressed in more technical and hence more specific and limited terminology.

If this be correct, it will cast light not only upon the question of integrity, but also upon that of the multiplicity previously discussed. To the former, I would address the following arguments: The endeavor to deal meaningfully with the knowable universe and to attain to knowledges both important to, and comprehensive of, such a subject is not one to be undertaken frivolously. And even if it be so undertaken, the frivolity would soon be discovered by students, readers, and other followers who would waste no more time in following further. Dullness in any measure can be forgiven, but woe to the frivolous who would philosophize; the lack of earnestness would soon be found out, but never forgiven. The survival of a man's philosophical reflections then is a testimony to the moral earnestness with which he undertook them; as they remain in the classic body of the literature we may become increasingly certain that he truly meant to explain meaningfully the realities that man can comprehend.

But how much more certain we can be of the intellectual competence with which he executed this earnest intention, if the results of his speculations continue to be studied, and remain available and of interest. For the very disagreement among philosophers makes many of them most ungenerous critics, and the selective process which determines the content of the history of philosophy is surely one of survival by most rigid criteria of fitness.

But at the same time we should be helped, by considering this matter of the nature of philosophic intentions and the earnest and competent zeal with which they are carried out, to understand the seriousness, and I think the ineradicability of the philosophic disagreement already noted. For when two philosophers disagree, their difference is not a slight or passing thing. What they disagree about—their common subject matter—is the whole of man's knowable universe. And the judgments articulating their disagreement are neither shallow nor narrow: they mean to express truths of utmost importance and scope.

IV

We are now ready to enquire more deeply and persistently into the question of the nature of philosophic thought, looking for an account of it which will be not only intrinsically tenable, but also helpful in explaining the characteristics already noted.

It has been contended above that philosophers attempt to formulate and express statements of maximum import and scope concerning the whole of reality, as far as this can be known or thought. It remains now to describe the methodology by which this philosophic end or goal is approached. Such a description must be of great generality or abstractness if it is to comprehend the declared diversities of method, which have differentiated philosophers and schools no less truly than diversities of conclusion. I think the following description approaches the necessary level of abstraction, and I can only hope that it will at the same time be admitted, by others who are engaged in philosophic enquiry, to be accurate and meaningful; I can say with honesty and confidence that it does articulate my conception of the methodology of my own reflections.

The philosopher *begins* with "all of experience" in whatever sense this is available to him. From his own most personal, innermost sensations and intuitions, ranging outward through the thoughts and awarenesses of other sentient beings, to the edges of the universe and their appropriate perspectives—this is all embraced within the ground or origin of philosophic thought, for to the extent that it is *present* to the philosopher it demands of him an explanation. "What everything is to everything" may be given as one definition of the totality of experience; and so far as this totality can be denoted by the philosopher's symbols, it will be found in his system, playing at least two roles: first, as already noted, it is that which his philosophic system will endeavor to explain, or "make sense of"; and second, it will be called upon to confirm or verify his speculations, when he is ready to claim success in his explanatory enterprise.

This second role bears a formal resemblance to the processes of verification found in the special sciences; but whereas a "fruitful" theory or explanatory hypothesis in a particular scientific area of concern must almost always have implications that allow predictions regarding areas of experience beyond those which occasioned the formulation of the hypothesis, there is in philosophic speculation no such advance or enlargement, since the entirety of experience has from the first been the "explained consequent" of which the philosophic system endeavors to be the "explanatory or explaining antecedent."

To recapitulate: the methodology of philosophy, considered most abstractly and simply so as to take general account of all historical diversities, has three major stages. First, the philosopher considers the totality of experience insofar as this may in any way be present to him, and regards it as something to be explained—as the consequent of an hypothesis of total scope for which he must now discover or conceive the best possible antecedent. Second, the philosopher essays the exposition, analysis and declaration of his philosophic system, with the intention and belief that this system will explain the totality of experience; the meaning of "explanation" in this context is simply this: If the whole-of-reality *is* as this system declares and describes it, then the totality of experience would be just as it is. But this explication of the meaning of explanation leads us directly to the third stage or moment of philosophic methodology, the confirmatory stage referred to above; it now becomes the philosopher's task to demonstrate the explanatory powers of his systematic account of reality, to justify his system by showing in comprehensive detail that it does indeed lead adequately and necessarily to the consequential account of the totality of experience.

Some might say, at first thought, that this third stage contains nothing new, being but the implicative or hypothetically conjoined enunciation of what the first two stages had already expressed. But what it does of course originally and newly comprise is the joining of these two preceding moments through their novel modal and causal qualifiers. "It is *because* the whole-of-reality is as described in this system, that the totality of experience *must be* what it is." This is the third-stage declaration of the confident philosopher. Even the modest and tentative exponent will say, "If . . . ., then it would be . . . .," and this is enough to support the claim.

V

The final aim of this essay is to suggest the nature of our proper attitude toward philosophies and philosophers. If I have understood the subject matter of the history of philosophy, it is an irreducible pluralism of explanatory hypotheses, in which there can be discovered broad and profound disagreement about the nature of the totality of experience requiring explanation; about the criteria of adequacy, both intrinsic and inferential, which any account of the whole of reality must satisfy; and about the nature of the implicative bond or verificatory demonstration by which the explanatory powers of the philosophic system are made evident.

Moreover, if I have given an admissible account of the philosophic enterprise and its methodology, then this disagreement seems quite understandable. Each of the three major stages admits of indefinitely great variety of specification; each particular specification can be held as an absolute tenet by its philosophic partisans; by self-willed right the philosopher need appeal to no higher court, and unprincipled compromise is equal anathema. The future of philosophy will witness and record the exfoliation and multiplication of systems, if history is any guide.

The implications for the student and teacher of philosophy seem clear. Really to study a philosophic system requires genuine "sympathy" with its exponents on all three methodological stages; it must be allowed by the student, and all who would aid him, that the originator or adherent of that system may be essentially correct in every particular. But this same sympathy or true tolerance must then be transferrable to another system, and on to all the others that the student would seriously study. Wherever it is withheld from a system, in any measure, repudiation and misunderstanding will follow in like measure. We know that this price is cheerfully paid by many; regarding many, most, or "all but one" of the systems of philosophy as vain or mad, they fasten tenaciously upon the narrow residue, without ever considering the wealth of genuine alternatives with seriousness or respect.

That this is their privilege as individuals, impelled by stronger forces than the desire to study philosophy, will not be denied. But that this attitude, in whatever degree, precludes in like degree the serious and successful study of the history of philosophy and the multiplicity of systems which comprise its content, I would most vigorously assert. The attitude appropriate to the study of philosophy must be one which makes that study truly possible: one that recognizes the irreducible plurality of philosophic systems and resolves to consider each system in its turn with the genuine tolerance of mind and will that its merit deserves and its understanding demands.

AN EXPLANATION OF PHILOSOPHY

By James K. Feibleman

I

WHAT IS PHILOSOPHY?

TO MOST people these days philosophy is a vague term with an uncertain meaning, having good though somewhat old-fashioned overtones but too confused, too irrelevant, and too mental, to be taken seriously. When not subordinated to religion, philosophy is thought to have become a merely academic subject, and finally science seems to have rendered it altogether unnecessary.

If such an answer is justified, then why are there still philosophers, why do men still write and teach philosophy? Are they merely foolish fellows who do not know that they are occupied with a dead field? Why is it that the physicists who were responsible for the revolution in physics all wrote philosophy; why is it that the classics in philosophy are so popular with the general reading public; why it is that as an academic subject philosophy is flourishing mightily?

Perhaps the last place to look for our answer is among the philosophers themselves, for they have no established opinions. It will have to be admitted that no statement could be made about philosophy that would be agreed upon by all philosophers, including this statement. Philosophers are seldom in accord, and this fact is well known; but what is not so well known is that this same situation can be found to hold also in many other broad human enterprises: in politics, in religion and in art, for example. Different types of government, rival religions, and various movements in art, have hardly been admired for their universal harmony. It is only in industry that cartels are formed, but the aim of business which is to make money is simple and plain, and anything but pretentious. The rule seems to be that the more fundamental the inquiry the more absolute its results; and the more speculative the question the more disastrous the practical application of the answer. Many disinterested persons have died miserable deaths in the attempt to throw open to controversy a problem regarded as settled. Have there been any wars worse than the wars of religion, say the wars between Roman Catholic and Protestant in sixteenth century Europe, or between Moslem and Hindu in our own day?

Look at the divergence in philosophies which is current at this very moment. On the continent of Europe, existentialism is the fashion and existentialism asserts that 'being' means 'being unwell,' in various degrees from visceral nausea to emotional anguish depending upon its advocates, with a resolution that the accompanying inability to face the making of choices should be cherished. In Soviet Russia, dialectical materialism is official, and philosophy as a consequence is founded upon the belief that all matter vacillates like reasoning and that therefore those who do the most material labor should rise to the top of society, preferably by violent means. In England and the United States, positivism is fashionable, and positivism recommends that all philosophy consists in pointing with pride to what the scientists do and otherwise in attacking metaphysics as so much nonsense. In the Roman Catholic Church, the philosophy of Aquinas is official doctrine, and it consists in reconciling with the combination of Plotinus and Christian revelation the newly recovered Aristotle of the thirteenth century, just as had been done for Islam by Averroes and for Judaism by Maimonides not many years earlier. In academic circles, philosophy is thought to consist in being grateful that when the history of philosophy accumulated and so provided a respectable profession for teachers, it incorporated the work of many men who wrote so ambiguously that constant reinterpretation is required. It should be added, perhaps, that each of these philosophies except the last has a rider which asserts that all other philosophies ought to be prohibited, or, if that is not enough, then that their advocates should be persecuted.

Admittedly, such one-sentence characterizations of contemporary philosophical positions are wholly inadequate, and are such as their adversaries might have given. Yet it is true that each philosophy always has all others as its opponents. The description is to some extent a caricature, but perhaps like caricatures brings out some of the more prominent and also more irritating features of the original. The intention, however, is to point to the extent of the divergence, for each of these philosophies has an enormous number of living adherents even though they have little else in common. Philosophy is a curious undertaking, and more striking when more closely inspected.

The philosophies described are those which have been advocated by the largest and most successful of institutions or which have won adherents by being the most admired. There are others of lesser fame which deserve at least a mention; for instance, American pragmatism which insists despite the best intentions of its founder that philosophy amounts to nothing except what it can lead us to do in a practical way. Many great philosophies of the past have sincere contemporary advocates who think that the best we can do is to bring their masters up to date. Plato, Aristotle, Spinoza, Kant and Hegel are among the favorites, but they are by no means the only ones. Serious philosophy ends somewhere after this, but the catalogue of philosophies includes much that is suspect or frowned on, and degenerates into personal attitudes and mystic cults, for each of these can count on some enthusiasts. The man who is not a success in the hurly-burly world is recommended to retire to the privacy of his study and there to take his defeat "philosophically," thus disclosing an understanding of philosophy as some sort of private consolation for public failure. An organization calling itself the Society of Master Metaphysicians existed some years ago in Philadelphia but was closed by the police for keeping the dead body of a girl, who had been a member, in the living room of a home until the neighbors complained.

Such a quick survey of the philosophical scene is quite sufficient to show what the chances are that any concert of opinion concerning the nature of philosophy could be reached. On the best interpretation it can be found that what Mr. Brown will allow (no matter who Mr. Brown is) is that what Mr. Blue is saying is after all only the Brownian philosophy bluedly — and hence confusedly — stated, inasmuch as the Brownian philosophy is the widest and truest that there could possibly be; and to this of course we could never get Mr. Blue to assent. Evidently, the broader the enterprise the less easy to define or even describe with any precision. As much difficulty would be encountered in the effort to define religion so that all religions could be included, and, though this is not so well understood, the same difficulties would arise in the definition of science. It would seem that the more universal the institution the less are we able to set limits for it or to put our fingers on precisely what it means.

The truth about philosophy is not easy to come by; for certainly if it were, men are not so arbitrarily obstinate that they would continue to disagree to the extent to which they have disagreed and still do. Those who are convinced that the truths they have discovered are not half-truths but the whole of truth are driven to such vociferous lengths as the imposition of their opinions by force; yet they would not have to do this if the evidence itself were compelling. But the failure of philosophers to agree has its good side, for one of the tasks of philosophy is the exploration of the field in which such opinions are relevant. We shall arrive at the truth sooner if we know what truth means, and we could find the true philosophy if we were able to look out for all the false ones. Perhaps this last task is endless and the truth impossible to approach through trial-and-error: a systematic survey of an infinite number of instances is by its very logic impossible. To conduct a survey, moreover, is not the sole task of philosophy though it is one. The more urgent requirement is that life, individual and social, hardly can be conducted except on a basis of some consistency, and consistency means applied (some would say implied) philosophy.

It is a paradox that abstract studies often make their greatest advances in those periods when they are supposed by their professional advocates to be utterly useless. Physics and mathematics were not cultivated for their practical value, nor brought to the pitch they reached by men eager to be of social service. The same can be said of philosophy. Evidently, the kind of intense and prolonged preoccupation that progress in such fields requires is possible only on the assumption that their applications do not exist. In this way are produced the theories which have the greatest concrete advantage. But the pendulum has swung so far that even the teachers of philosophy can think of no value that philosophy can be to men of affairs. And here they are wrong, as the study of the effects of the history of philosophy and the subject-matter of the philosophy of culture could tell them. If indeed there was no more to philosophy than its textbooks and its classrooms, the ambitious students were well advised to stay far away from it. But this is not the case.

Under such baffling circumstances what, then, are we to say that philosophy is? It might be obvious supererogation to say anything. And yet the same urge that leads a man to become a philosopher must drive him to assert his own interpretation of philosophy. The preferred opinion is that what philosophy ought to be (though what it has been on only the rarest of occasions, perhaps solely in Socrates' day) is an enterprise to lead all those who suffer from philosophical opinion so confirmed that they would impose it as the absolute truth upon their neighbors, into the more passive and safe channels of unsettled speculation.

Philosophy has been engaged in freely only by the ancient Greeks, who may have invented it, and not again until our own times. Otherwise it has been the slave of other institutions, religions chiefly. Al-Ghazzali, who wrote the *Destruction of the Philosophers,* succeeded in having it officially abolished. But three centuries later he was followed by Averroes, one of the greatest if not the greatest of Muslim philosophers. Philosophy has shown a toughness and a persistence which indicates the presence of something powerful. It is now threatened again, this time by the philosophers themselves, who wish to subordinate it to science. We would do well to put the analysis of the question of what philosophy is ahead of the statement of any attempted answer. Perhaps philosophy at its best is the method whereby we persuade both ourselves and others that in such fundamental matters the analysis of questions is to be preferred to the imposition of answers.

II

An Analysis of the Question

The analysis of the question, what is philosophy? can be made in as many different ways as there might be answers proposed for it. Each philosopher, however, brings to his task something of the same ambition and equipment, and, if he is successful, achieves a similiar result. Let us begin by looking quickly at what this approach means.

There is perhaps no one living who does not feel about life in addition to much fulfillment some inadequacy. Every comparison of similar things suggests some imperfection: deer run faster than we can, sunny days are better for plants than cloudy ones, some men survive much longer than others and are healthier. For some imperfections our efforts have been helpful; we have improved farm breeds, we have altered our environment in our own favor, we have learned to extend the span of human life. But many more difficulties remain, not only with our environment but also with our outlook. Individual desires frequently clash; the actions of many of us, if not of all, are in conflict, and we ourselves are among the imperfections. Between individuals lying, cheating and even murder, are common enough, and broader social conflicts are even worse, for war is still employed as a means of settling disputes.

All around us, then, and also within us there is that sense of imperfection, of limitation, of problems settled without the use of reason and often even in disregard of the facts. When we are able for a moment to get some distance between us and the immediate crisis, whatever it may be, we are able to see that our behavior and perhaps also the behavior of our contemporaries both individually and socially has been to some extent a piecemeal affair. We can note then that our actions and theirs are to a large measure self-defeating. But somehow there must be an understanding of the whole of existence into which the separate parts fit and can find themselves, if only we were able to grasp existence as a whole. At least we do see the old primitive questions, 'why are we here?' and 'what should we do?" as separate and subordinate parts of a much larger question, 'what is there?'

The resultant enterprise is the one we call philosophy. It consists in the attempt to fit the various parts of our activity together into a meaningful set by way of an understanding of the whole. And it tends to see this whole in terms of perfection. Philosophy begins with a prejudice, a prejudice in favor of the excellent, the detached, the ultimate. The ambition of philosophy is understanding.

It might be objected that understanding is not confined to philosophy and that therefore it cannot be used as the equivalent of philosophy. After all, has not religion sought understanding as well, and has not art done so, and science, and even common sense? The list of enterprises which have understanding as their aim is even wider than those we have listed. Every practical undertaking, such as engineering, and political or economic systems, has sought for an understanding, too; and so even have such humbler technologies as cooking and farming.

The objection is sustained, and may help us in our exposition. Philosophy is both a separate and a collective attempt at understanding. For it consists in the presuppositions with which each of these manifold enterprises begin. That such presuppositions may be silent and unacknowledged makes no difference to their existence or their presence. Philosophy exists separately in every other enterprise also as the structure of its method. Every enterprise has some consistent method, and that method can be revealed by means of the analysis of its logic. Finally, every enterprise has conclusions which stand in need of interpretation, and philosophy can often furnish that interpretation. Thus philosophy, implicit or explicit, exists at the beginning of an enterprise as its presuppositions, is present during its workings as the logical structure of its method, and finally is required by its interpretation.

Such a claim could stand examples. Let us choose for this purpose one of the more ambitious enterprises and one of the humbler.

The ambitious one could well be that of religion. Religion usually begins with the insights of some individual, who may or may not claim divine revelation for his utterances. But no religion can secure the allegiance of many persons without becoming established as an institution. And to do so it must adopt a philosophy, either openly or in some unpromulgated fashion. The consequent theology has its presuppositions, its method and its conclusions. And if these are to be respectively defended, deepened and interpreted, considerable philosophical acumen is required.

Suppose we take as one of the less ambitious enterprises a code of law as adopted in some country. Chaos and confusion, social disaster, result if the law is abandoned and no other substituted; so it is necessary to social order, and is vigorously defended; it is, in other words, established. But is it all that a law ought to be? On what does it rest? How does it work? What can we conclude as to its meaning? These are the same questions we have been asking, and their answers call for considerable philosophy.

Thus in the separate instances of enterprises other than philosophy which seek understanding, philosophy is implicitly involved and explicitly needed. But this is not all that there is to philosophy. For this is merely philosophy in the service of other enterprises; what about philosophy itself as an enterprise? We have asserted already that philosophy is both a separate and a collective attempt at understanding. Let us see what each of these means.

Philosophy engages in criticism and in system-building. It is the critic of presuppositions. Presuppositions although first in systems are still parts of systems. When they are included within systems, we call them axioms. The history of the opinions regarding the status of axioms would make an interesting book. The Greeks held them to be self-evidently true, and supposed that every axiom carried on the face of it, in addition to whatever else it asserted, the assertion of its own truth, and was thus two assertions rather than one. Modern philosophy has decided that truth is irrelevant to axioms, and that we have no way of knowing and little reason for caring whether they are true or not. What we do care about is the validity and applicability of the deductions from them.

This judgment too may be changed in time. For if the axioms are not true and we make valid deductions from them and then apply those deductions, must not our applications eventually fail? A popular argument of the nineteen twenties was that because Mussolini had the trains running on time in Italy, fascism worked; but it was not many years before he ended hanging by his heels, and fascism lay in ruins. Democracy, it seems, has worked for longer. Application is not a final proof of the truth of principles, only a strong argument in favor of them; but we must be sure that we have given application a long enough run of instances to enable it to work itself out somewhat. But it seems clear that if we apply principles many times and under diverse conditions successfully, does this not argue something for the truth of the axioms from which such highly applicable principles were deduced? In any case, philosophy is heavily involved with the axioms and with the question of truth.

The axioms of any enterprise in addition to being considered inside that enterprise, as we have been doing, may be considered also outside it. We may suppose that the axioms were themselves deductions from some more abstract system, and then seek the outlines of such a system. We may look, in other words, in the direction of ultimate axioms. At this point our argument is joined by two other considerations. Is there such a thing as an ultimate set of axioms which can lead by deduction to all of the axioms that we know from other enterprises? And does all philosophy have a single set of axioms? Finally, are these axiom-sets one and the same?

The last few problems are not easy of solution, perhaps they are even incapable of it; certainly even as problems they are not quite so simple as they have been stated. But something definitely lies in that direction, and that is why philosophy at times seems to be so simple: it is seeking for an ultimate simplicity, one that has no assumptions of its own because it underlies all other assumptions. Philosophy as the critic of presuppositions finally has its work cut out for it.

Philosophy is not only critical but engages also in system-building. It must put together the findings of all other enterprises by means of equipment especially designed for this purpose, and in this way seek for a general understanding. It must make up a whole explanation out of parts such that the explanation in any other enterprise or undertaking would be some part of it. Obviously, such a task, like the other, is complex and endless. Something lies in this direction, too, but far away and difficult of access. What do science and art have to do with each other? To find the answer to this question, it would be necessary first to have a precise understanding of what both science and art are, and then to posit some very much larger structure in which they were parts among many others. It is as though a man were to have in one hand the faucet of a washstand and in the other a tile from a roof. Could he from such elements ever imagine the form and function of a house? And yet that may be the task of philosophy: impossible on the face of it, and yet demanding of at least a surmise. For if we have not the equipment to make answers to such questions, it is also true that we lack the ability to refrain from asking them.

It is true of all vigorous periods such as our own when many things are being changed — the sciences proliferating at a fantastic rate, religions being revived and reconditioned, the arts putting forth their claims to existence in fresh ways — that the individual whose most urgent task is the gaining of a living may feel intensely the need to understand even what he does not know. For him at present all lines go away from the center: sex for the Freudian competes with cash for the Marxist and revelation for the Christian as the sole explanation and cause of events; and the politician, the priest and many others separately claim exclusive ascendency over him. What is he to think about things in such a world? He can keep busy and try to put aside the question, but when he does so he suppresses a craving to know how things are, which is as fundamental as his other cravings to eat and to love.

III

WHAT DOES PHILOSOPHY TRY TO DO?

So much for the ambition of philosophy; and when we try to analyze the question, 'what is philosophy?' we have seen that we must include first and foremost its aims: what does it try to do? Now we must turn to the second portion of our analysis of the question, and ask ourselves, in what way does philosophy try to achieve its aims, what equipment does it bring to its task?

Every enterprise has its own technical tools, including in most cases its own special vocabulary, and philosophy is no exception to this rule. Here is where one of the most prevalent difficulties comes in. For no one would try to understand mathematical physics or biochemistry without some special training, yet everyone feels quite sure that any philosophy which is not readily available to the average educated person is somehow inadequate: it must be wanting in something as a philosophy or be badly written. The average man blames himself for his failure to understand physics, but he blames philosophy for his failure to understand philosophy. And yet philosophy is no less technical than physics or any other special branch of learning.

Philosophy is divided into a number of special areas, technical divisions for which there are names. Among these are: logic, metaphysics, epistemology, ethics and aesthetics. Every comprehensive philosophy contains all five divisions or branches, but most favor some one over the others. It will be best perhaps to glance separately at each of these.

Logic is the theory of abstract systems. These are usually approached deductively, though there are other approaches which would have to be included, such as the inductive. A system is a structure in which some propositions follow of necessity from others. The extent to which this demand is met is the measure of their consistency, and the number of such propositions — their inclusiveness — is a measure of the system's completeness. A good example is mathematics. Indeed mathematics has been described as "all abstract deductive systems." The terms in which such propositions are framed are usually those invented for the purpose and therefore have no second meaning which could lead to confusion. But logic has been traditionally worked in ordinary language and of course to some extent still can be done in that way. Logic leads to mathematics, but that is not the only application. Applied logic leads also to any orderly or systematic material.

One important area in which applied logic can be of great service is human reasoning. Logical reasoning is of course correct thinking. In the nineteenth century, logic and reasoning were regarded as so intimately connected that logic was supposed to have been derived from reasoning, and was called the laws of thought; but more recently we have returned to the view held by the ancient Greeks, according to which reasoning is only one place where logic can be applied.

We turn next to metaphysics, which is a system of ideas wider than any other existing system. This means variously the construction of first principles or the criticism of assumptions. That legal systems presuppose ethical principles was a metaphysical discovery, as was the proof that some philosophies are insufficiently inclusive. That metaphysical criticisms are made slowly and that such opinions accrue, have blinded men to the fact that they were made by metaphysics at all. The process of showing that there is no safe and rigid bottom to any discipline is a tedious and unpopular enterprise, especially to those who have to work on the assumption that there is. They have been led out of the difficulty by some professional philosophers who have contended that metaphysics is reducible to the language in which it is expressed, and, therefore, since it is not about anything else, clearly it is not about anything.

Philosophy in its constructive or systematic phase is known also as ontology. Generally speaking, ontology is positive and constructive metaphysics; metaphysics, negative or destructive ontology, although it should be remembered that criticism is analytical chiefly and so not always destructive.

The propositions of metaphysics are combinations of terms. The terms themselves are known as categories; these are principal classifications, and set the entire determination for a philosophy. Once we have chosen our terms, we have more or less decided what the philosophy which will be made up of them is to be. Familiar examples of such classifications are: 'essence and existence,' 'form and matter,' 'mind and matter,' 'possibility, actuality and necessity,' etc. The categories of classification are in effect basic attempts at explanation. To employ a single category has the advantage of simplicity, but it explains little. Two seem to do more, and three more still; there are examples of more than three but they are rare: a fourth in some way repeats what some one of the first three has done.

The relation between logic and metaphysics is a close and intimate one, although the two studies are kept quite separate and often are engaged in by specialists who have little or no regard for each other's field. Most logicians are not concerned with metaphysics, though there are important exceptions, but less often the metaphysicians are not concerned with logic. Perhaps it would be better to say, the metaphysicians are less concerned. A logician is a man who takes his metaphysics for granted, that is, he runs it in as a set of silent and unacknowledged assumptions. A metaphysician of any scope is one who freely acknowledges the logic which his metaphysics assumes and employs. Hegel's logic was important to his metaphysics, even crucial to it, as he knew; but the metaphysics underlying Boole's logic was never mentioned, since the logic can be set up and operated without explicit reference to any metaphysics.

Most of the great philosophers regarded metaphysics as central to philosophy, but not all. Some have preferred to start with epistemology and then to derive their metaphysics from it. Epistemology is the theory of knowledge, it studies how knowledge — including metaphysical knowledge — is possible. How do we know, how reliable is our method for obtaining knowledge? Epistemologists argue that to start with metaphysics is to take a certain philosophical method for granted, to proceed naively. Metaphysicians reply that all epistemology and indeed all methodology (since that is what it is) always has its hidden metaphysical presuppositions without which it could not examine the method of philosophy, that epistemology is merely a sort of examination of the validity of metaphysical proofs, and can therefore not take place until there is some metaphysics to examine. Kant is the chief exponent of those philosophers who held that epistemology precedes metaphysics.

But in any case, there are the three central studies in philosophy: logic, metaphysics and epistemology, and any philosophy which leaves out one of the three is the poorer for it. And a poor philosophy is hardly a philosophy, since it lacks the ingredient of inclusiveness which is more vital to philosophy than to other studies. It is currently fashionable in England and the United States to omit metaphysics from the trilogy, on the grounds that it has no object and that therefore it refers merely to the language in which it is expressed: that metaphysics is about metaphysics. But it has been pointed out (though not to the satisfaction of those who advocate such a position) that the metaphysical assumptions of those who claim that metaphysics has no object are great indeed, and that therefore the argument against metaphysics is self-refuting. If the very first statement made in this paper is correct, then there are those philosophers also who would discontinue philosophy as a viable enterprise, and turn its advocates toward other fields especially to mathematics, physical science and psychiatry: to mathematics for the logic underlying metaphysics, to physical science for having the exact method of reliable knowledge which has antiquated metaphysics, and to psychiatry for treating those who suffer from the mental illness which allows them to have the illusion that there is such a thing as metaphysics.

So much, then, for logic, metaphysics and epistemology. There still remains two divisions of philosophy to be examined: ethics and aesthetics. These along with theology are the so-called value studies.

Ethics is the theory of the value or quality known as the good. Like all values, it cannot be described to those who have not felt it, and is thus quite different from rational ideas which certainly can be explained to those who have not known them. Like the color, red, or the taste of sugar, the quality of the good cannot b conveyed to those who have not seen red or tasted sugar. But we can talk about it, and hope that the common experiences upon which we are relying for our meaning will be linked, and in this way communication will take place. In past times (and still among many today) the good is held to be exclusively human: the good is what is good for man; or even exclusively subjective: the good is what is good for me. But as in the case of logic that is not the Greek view, and it is not the most modern one, either. For if we insist that whatever preserves any organization whatsoever, whether a stone, a tree or a man, is good for it, and whatever destroys it is bad for it, then the good is simply the quality of the bond between wholes whatever those wholes may be.

An organization is a system of parts made up into a whole. Some organizations are loose collections of parts, some are very tight ones indeed. Aesthetics, which is the theory of that value or quality known as the beautiful suffers from the same failure, or, at the very least, the same difficulty in communication that we have already found to be the case with ethics. What is beautiful can be pointed to in the hope that its beauty will be shared, but it cannot be communicated directly any more than any other value can.

The theory of art is a theory of the beautiful, and the connection between art and beauty has been so close traditionally that men have supposed beauty to be exclusively human, for after all is not art a man-made affair? But once again, as in the case of logic or of ethics, there are those who maintain that art is not man-made exactly but rather man-discovered and could not be brought into existence if the values that are contained in a work of art had not already existed in some sense as possibilities. The sounds of which a piano sonata are composed already existed as sounds, and their combination in the sonata was already a possibility, even though it required the intuitions of a genius to see it. But intuition is after all just what it is often called: insight, and where there is nothing to be seen it cannot occur.

Aesthetics, then, can be considered as the degree of perfection with which the parts of any organization are fitted into the whole without remainder, and so wherever there is an organization there can be beauty, whether anyone is there to appreciate the beauty or not. Appreciation is something else again. If beauty lay in the the eye of the beholder, it would be possible for every connoisseur to be his own artist. Unfortunately, perhaps, this is not the case, and appreciation can only begin where the production of art leaves off; and so they are by no means the same.

In aesthetics, then, we are once again close to the ancient Greek view of things. It is easy to see why so many philosophers hold the Greeks in high esteem, for they pioneered many of the philosophical pathways, and opened up fields in a fertile way to speculation; so well, in fact, that we are still exploring many of them today. Our philosophies however fashionable and *avant-garde* often prove to be assemblies of ideas which had been discovered by the Greeks who did not, however, assemble them in just these ways.

For the Greeks, too, ethics and aesthetics lay close together. The perfection of organization is not too far from the relations between organizations; the one affects the other. There are other studies in which values are explored besides those of ethics and aesthetics: theology, for instance, the theory of the holy. But speculative philosophy has usually had to turn its conclusions in this field over to other institutions, more specifically to churches, in which they became official and hence obstacles to further speculation, except what could be conducted in the special terms of the theology adopted. Institutions which endorse particular philosophies do explore them with an intensity they would not otherwise receive, but they put an end to the consideration of philosophies which may be rivals to the one endorsed.

IV

How Can Philosophy Help the Untrained Individual?

We began our inquiry by noting that the discord of contemporary voices in philosophy discouraged any hope that philosophers might some day agree upon the meaning of philosophy. In the next section, therefore, we decided that the answers in philosophy were, philosophically speaking, not as important as the questions. And so we decided to explore the primary question, what is philosophy itself? It was found there that philosophy can be explained chiefly in terms of its aims, and that these aims, roughly speaking, reduce to a single one: philosophy seeks the most general kind of understanding. It then became necessary to learn what tools philosophy uses in its efforts to satisfy its aims at understanding, and this called for a breakdown and description of the various divisions of philosophy.

In all three sections of the foregoing exposition, we were concerned with *theoretical* philosophy, with philosophy in search of the truth. Its aim was discovery. We are now to turn to a second broad division which is concerned with practical or *applied* philosophy. As usual in philosophy, we are in another large area of disagreement. Some philosophers would shudder at the thought that philosophy has its useful side, and it must be admitted that they too have made important contributions to philosophy. But here we shall try to develop the theme that philosophy is of the utmost usefulness.

Philosophy can be helpful in various ways to the individual and to society. Let us consider first those which are of use to the individual. It should be added quickly that we are not here talking about the professional individual merely. Our discussion is directed toward the individual who has no special technical training in philosophy.

For the average individual who has other things to do and to worry about, it is not at all easy to see what any of this has to do with him. The businessman, for instance, spends most of his waking hours in getting a living, and what time he has left over is usually devoted to his family, to friends and neighbors and their interests, and to keeping up with what is going on in the world. In a period such as our own, when as a result of wars, of social conflicts and of new developments in knowledge, all the old beliefs have come unstuck, the individual feels lost and in need of help in fighting his own sense of insecurity. The more he learns about psychoanalysis, Marxism, the new physics, the chemical industry, the rise of mathematics, automation, the ballistic missile, the revival of religion, the dangers of tobacco, the overproduction of automobiles, the spread of television, rock-and-roll, the vast political parties, and the influence of oil producers, the more confused he becomes. Behind him lie the old social ambitions, ahead there is nowhere to go and nothing to do. How possibly can philosophy be fitted into such a picture?

The answer is not a simple one but there is after all an answer. For we have not been altogether fair to our businessman. He is not just a doing machine, he is also a thinking and feeling person, with all the full dignity of his perquisites, and these functions are not exhausted by his role as business partner, husband, father and citizen, however important each of these may be. True, he may go to church on Sunday there to have all his doubts and questions assuaged in the comfort of a faith; but there is more to it than that. For, although pious, he is nevertheless a separate person and to some extent different from his fellows. Now, every man in becoming a person develops a point of view of his own, overlapping with others to some extent, but also to some extent unique. There are times when he would like to know what that point of view is, and occasions upon which he longs to make the acquaintance of his own consistency. If we say here that philosophy is in a position to help him, he can answer that this remains to be shown. In the last analysis, he will have to perform this service for himself; but philosophy can furnish him with the tools.

What can philosophy do for the individual? We have listed five of the branches or subdivisions of philosophy; these are: logic, metaphysics, epistemology, ethics and aesthetics. There are accordingly five tasks it can perform. Before discussing them, it will be helpful to mention a sixth with which it is always well to begin. This sixth is the history of philosophy itself.

The history of philosophy is an account of what some of those who have given their lives to the study of philosophy have thought about it. It is, in short, a sampling of systems, assumptions and insights. A reading of some good history, therefore, will acquaint the uninstructed individual with the sort of thing that philosophy discusses. It will provide him with the experience which will make it possible for him to recognize the kind of fundamental belief or statement which is philosophical in character. For when he comes to his first task, which is the exploration and discovery of what he himself believes, he will be equipped for it.

A man's beliefs consist in all shapes and varieties of suppositions, some significant, others trival; some profound, others superficial; some known to himself, others entirely unknown; some gained through systematic learning, others acquired through chance experience; some serving as a basis for deduction, others deduced. Is it true, for instance, that we ought to put our faith in what exists now, and gives less credence to what has passed or to what is to come? Have we not perhaps transferred our belief in substantiality of matter to those documents which relate to its possession? Do we not, for instance, slip into supposing that the United States Steel Corporation *is* the shares we hold in it? Is the evidence for immortality sufficient to justify the sacrifices we make in this world to gain it? Are the physical laws that cannot be evaded any less real than the pieces of matter we see answering to those laws? For the most part, it is fair to say, we are not well acquainted with the most fundamental of our beliefs; we are not in the habit of searching for them but we are in the habit of acting from deductions made from them. In short, it is our practice to do what we believe and to have our beliefs show in our actions, without bothering to examine, in minute detail and the fulness of consciousness, what those beliefs are and how strongly they are supported.

To most of us it would come as a great surprise were we suddenly to be confronted with an array of our beliefs. We should not recognize all of them, and we should shrink with horror from some. If we were to argue backwards from our actions to the beliefs from which such actions would follow as deductions, we would emerge into the presence of some of our beliefs, but there would still be others which were not uncovered by such a method because they had not yet had the opportunity of leading to actions. Not all beliefs are called into overt behavior; and those that are, depend for their initiation upon some event in the world of external events to trigger them.

The task of uncovering fundamental beliefs is not a simple affair nor one easy to achieve, but it will be aided if we know the sort of thing we ought to be on the lookout for. The history of philosophy can be helpful here. The sixth way in which philosophy can be useful to the individual — and the one we have chosen to mention first — is the study of the history of philosophy, for it can make the individual familiar with the sort of fundamental beliefs he is likely to find in himself when he examines what he has taken for granted and what his feelings and actions reveal. If he has made something of a study of the history of philosophy, then he is familiar with the kind of world philosophers live in, the sort of ideas they advocate, and the variety of conflicts their rival ideas engender. In the examination of the other five ways we shall have to assume that he knows what his fundamental beliefs are, although this is quite a large assumption.

Let us assume it, anyway, for the purposes of getting on with our explanation. We have, then, an individual and his fundamental beliefs. And now we are to show him how he could apply to them the material which is contained in the five standard branches of philosophy.

Logic was first, and logic has to do with consistency and completeness.

To what extent is it possible to discover among anyone's fundamental beliefs a consistent system? Obviously, a certain logical approach to the problem is to be desired. The beliefs will have to be sorted out, and those that are the more primitive and that perhaps lead to some of the others must be set aside as axioms. In other words, if possible a system will have to be found among the beliefs.

It is not always possible to discover such a system, of course, for a system is not always present. In that case, it would be advisable to fall back upon a secondary type of procedure; and then the individual would have to be satisfied if he could find among his beliefs some which are consistent. If belief A is consistent with belief B, and if belief C agrees with belief D, we shall have to forego the fact that A and B taken together are not consistent with C and D taken together. We have, then, instead of one consistent system two separate consistent systems, each of which is somewhat lesser in scope.

We are setting for the individual arduous tasks. He has to ask himself whether his system of fundamental beliefs is well-supported. The first test of support lies in the principle of consistency. The second and third tests lie in showing the system to be allowed by truth and fact. Does the individual find that his system of beliefs contradicts what in other systems is known to be true? Does it conflict with crucial facts? Then he will learn that it is unsound, and that some revisions are called for in it.

A brief glance at the history of such matters ought to shake us in our facile certainty that what we believe to be is. Public and widespread support for our beliefs is comfortably confirmatory. Yet look at the size of the mistakes which have been made. Astrology had a longer run than almost any other human discipline, yet it is wrong, and astronomy did not come to supplant it for thousands of years. The same statements can be made about alchemy and its replacement by chemistry. There is a simplicity about false knowledge which renders it attractive; numerology can be mastered a lot more quickly than applied mathematics as we have it today. Then, too, there is the charm of the mystery which false knowledge always carries and which the truth endeavors to dissipate. The magic and the fear of the unknown is more bewitching than the complexities of the known. Yet in addition to the ease of belief and perhaps underlying it after all, men have a keen desire for the truth, and that alone may save them. It is logic and logic alone which can be counted on in this extremity, and logic is nothing more than the cry for two kinds of consistency: consistency with already established generalizations and with fact.

The criterion of consistency seems at first glance to be a most awfully strict and rigid one. Why, it will be asked, can we not allow the individual more latitude in his beliefs? Is it not almost inhuman to be so monstrously logical? Certainly an individual who always behaves in a narrowly correct way always strikes others as insufferable. There is some strength to this objection; still, let us examine it.

Beliefs, we have seen, do not end there: they lead to feelings and they lead to actions. It is their actions here with which we are for the moment chiefly concerned. When two beliefs which are contradictory both lead to actions, the actions may either cancel each other out or bring about considerable conflict. Suppose a man believed that charity is good, and also that the giving of alms is weakening and therefore an injustice to those to whom it is given. If he were to act from both his beliefs, he would give away what he could spare, and then take back what he had given. If he believed that all wars are wrong and none justifiable, and also that a man owed his duty and perhaps even his life to his country, then the conflict would be so powerful within him that it could lead to neurosis.

But to return, the advantage of knowing what your beliefs are is that you can then find out whether they are true, and, if they are not true, exchange them for others that are. The second step is to match them against the relevant knowledge, which is for good reasons commonly accepted.

Consistency, as we stated at the outset of our discussion of how logic could aid the individual with his beliefs, is not the only requirement. There is also completeness. Consistency demands that all of the elements within a system fit together; completeness demands that all of the elements that ought to be included in a system are in it. It may happen that beliefs are consistent yet wanting in scope, they may be too narrow. It is possible to believe in very little of anything and to have that little consistent, and yet to find such beliefs inadequate. Thus the second test of beliefs has to do with their completeness: we must ask that the beliefs of the individual be both consistent and complete. Thus we shall have to ask him to stretch his beliefs over a wider area than they had hitherto covered. The well-informed man has opinions concerning a vast number of topics. But those who know little have fewer occasions on which to hold false opinions. The man who does not know the oddness of the behavior of materials under enormous speeds or high pressures is in no position to entertain false beliefs concerning the causes of such behavior. In order to discover whether his beliefs are complete even if they have proved to be consistent, we shall have to see to it that he has the proper amount of knowledge. Yet who knows enough? The world is always bigger than our knowledge of it, and the processes of gaining and testing knowledge will have to be a continuing one.

Beliefs are not static affairs; they lead to feelings and actions, and they may also lead to thoughts. We think, feel and act in accordance with what we believe; there are few who are able to believe in one way and behave altogether in another, and they are so divided as to require the services of a psychiatrist, for such split personalities are of the schizoid type. The confusion in most people is of a more modest and partial nature. At both extremes there seems to lie great peril: we may be monsters of single-mindedness or monsters of dual personality. We strive to move away from inner conflict and toward consistency, without the fear that our logicality will reach such a pitch of proportions that we will become a menace. For the single-minded man may be a prophet as well as a sinner, a Socrates as well as a Hitler; but there need be little concern that most of us are capable of so much good or evil. The earnest content of the usual thought, feeling and action betrays a want of profound belief rather than a surfeit. To believe with a breadth or an intensity sufficient to lead to extensive thoughts, feelings or actions, is given to the few: to the mathematician in the case of thought, to the artist in the case of feeling and to the political leader or the saint in the case of action. For most, it is a case of poverty of insight and conviction rather than the reverse.

The third and last criterion then, by which the individual's beliefs must be tested is one that we may call deducibility. Another name for it, and perhaps a more descriptive one, would be fruitfulness. It is one of the tests of a set of beliefs to ask of it, how well and easily does it lead to thoughts, to feelings and to actions? A philosophy for the individual means a set of beliefs held so deeply that he does not know he holds it and yet one strong enough to influence all his thoughts, feelings and actions. So much, logic can do for the individual in the analysis and improvement of his fundamental beliefs. We can now recognize in these the outlines of his philosophy.

We turn next to the contribution of epistemology. How does he know that his beliefs are sound even after they have met the logical criterion of consistency? To answer this question, it will be necessary to examine the evidence. What is its character and how strong is it? There are many tenuous threads to follow in this connection; the theory of knowledge is not so precise an affair as logic and there are no absolutes like contradiction to be found. We shall be interested at this point for the individual in an inquiry on how his beliefs were acquired, on what they rest, and on how tenaciously he holds them.

His beliefs are so much knowledge; but how did he acquire this knowledge? The internal workings of the process of belief belong to psychology. The psychological process of the acquisition of beliefs has been little studied. We hear a great deal about learning and concept formation but hardly anything about the fate of what has been learned. Learning is not necessarily belief. I may learn that the importance of theoretical science is not understood by the members of the United States Chamber of Commerce but I do not have to believe it. The subtle ways in which we come to believe what we learn await perhaps a further knowledge of the workings of the nervous system.

For the present we are confined to saying that the ways in which beliefs are acquired can be traced to their external origins. There are many obvious sources: tradition, habit, sense experience, reasoning, acting, feeling, and what else not. In addition, other sources exist which are less obvious, and it is probable that any of the experiences of our whole life could, and many in fact do, contribute to the sum of our beliefs. The process is not a visible one nor do we necessarily know when it is happening. But there must be a point after we have acquired the understanding of a statement at which we accept or reject it. We may do both in varying degrees. For the most part, with rejection we have done with the statement; but not so with acceptance, for thereafter it will be part and parcel of ourselves. A statement we have come to believe may lie dormant if circumstances permit, and so for all practical purposes cease to exist, or it may call us into decisive action when relevant events arouse us to the awareness of its presence in us.

Were the sources of our beliefs such that we would do well to accept them as profoundly as we do? Are the beliefs themselves as sound as we take them to be? We do not know, and usually we take no pains to find out. Each belief would have to be examined and weighed, for again many beliefs conflict and not all tell the same story. There are some things we grow up 'always having known' that just are not so. As we imbibe the 'common sense' of our parents' generation and our own, we little realize that it consists in an ancient and inherited metaphysics. It is a jumble of stuff, some true and some patently false. Rain on Friday by no meteorological knowledge that we possess could possibly 'cause' rain on Sunday. The healthy and wealthy are not always those who go to bed early and get up early; and so on. But on the other hand, two plus two do make four. It is not the knowledge that we know we possess and are well able to defend that is of interest here, but that other knowledge that we so take for granted that we could not imagine its being questioned at all. It will often turn out that the knowledge we regard as the most secure and unassailable is in fact the most unsupported and shaky. The discipline of epistemology is a training in the seeking out and detecting of false knowledge lurking in our beliefs.

The individual will discover at this stage in the proceedings that he has a strong tendency to continue to believe what he has always believed, that the ideas he accepts tend to persist in him. After all, he has allowed them to become deeply imbedded; they have dictated his actions, and he has come to live by them. How, then, can he give them up? Not by any mere act of will, surely. Beliefs are not acquired nor abandoned except upon sufficient evidence. We are so constituted as to believe what there seems, to us at least, to be good reason to believe, and to doubt only where there seems good reason to doubt. To be asked to believe otherwise, appears to us litle short of insane. Our sanity depends upon taking certain beliefs as positive knowledge, and in putting them into practice, usually along with our fellows. We act usually in concert with our fellows or else in individual ways which they themselves would approve. A basic philosophy exists in every society, and is accepted so profoundly that its very existence is hardly suspected. Reason to believe based on evidence for belief, and reason to doubt based on evidence for doubt, is the only reliable starting point for arriving at knowledge. But such reasons are often complicated and subtle, and therefore hard to come by. We need to learn something of their properties and characteristic hiding-places, if we are to learn about how we are able to maintain our most fundamental beliefs; and we need to do this if we are to learn about ourselves.

Epistemology, then, is the study whereby we acquire the method of putting our house of knowledge in order. Even when we succeed in doing this, if we ever do at all, it is not a final or static condition. The task of mounting sentry at the gates to belief is a never-ending one. We must be continually on our guard against accepting too easily as knowledge the statements which, because they may be fashionable, clamor for admission. It is not sufficient to believe in the truth of an idea simply because there is not on the face of it any reason for doubt. There must be to the contrary positive evidence for belief. To doubt that a statement has come equipped with evidence sufficient to justify belief in it does not mean that thereby doubt is required. We believe what there is evidence to justify believing or to justify doubting, and otherwise we suspend judgment. We should never doubt simply because there is insufficient evidence for belief. Otherwise, all agnostics would be atheists, which they most assuredly are not.

We are pledged to the question, how can philosophy be of help to the untrained individual? And we are endeavoring to suggest answers to him in terms of the main divisions of philosophy. With logic as the theory of assumptions and deductions — the theory of systems — and with epistemology as the theory of evidence — the theory of how knowledge is possible — we have now methodologically equipped our individual so that he is able to take a wider view. That wider view is the one named ontology.

We are purposely omitting metaphysics here, because metaphysics is critical in the negative sense, and it is presumed that our individual has already been through that phase when he considered rival systems as he was making their acquaintance in the history of philosophy. Ontology is positive and constructive, and that is what he is prepared for first and indeed what he needs now. We shall postpone metaphysics for him, then, until after he has dealt with ontology. An ontology is a system of ideas sufficiently wide to include existing knowledge. Logic examines its nature in so far as it is a system, and epistemology examines its acquisition and retention. But in the study of ontology itself, so to speak, we have to do with the quality of its content. By the time we ask the individual to appraise his ontology, it has already been formed, oftentimes largely without his conscious planning or even help. But now we are on the threshold of an area where we need such help. For were conscious processes able to do no more than unconscious processes, consciousness itself would be suspect. The picture of the whole man is one which includes his most profound opinions, and these, we shall assume, are more systematic than unsystematic. They were formed by many influences, they were fed from many sources; information has poured in upon them. They came from the experimental sciences but not only from them, for the world is larger than any specialty, and includes what art has to contribute and what other, and lesser, disciplines can add. In the sum, this is much. Being a whole man and seeing as one, the individual must also be convinced as one; also, what he selects to see is determined to some extent by what he knows, feels and does, and so can be traced back to what he believes. This synoptic view cannot be formed in the first instance nor held together synthetically; it must be arrived at by an act of insight. Now, the history of philosophy has already shown our individual something of what such insights look like after they have been formed, verbalized, and exposed to the light and the years. He knows a little now about how to make up his own ontology, or, if he cannot, then he has learned perhaps which of the classic alternatives to adopt.

A whole man, his equipment, his peculiar perspective, his aims and ambitions, his degree of perseverance, all center upon his ontology and are immensely affected by it. Can he then afford to ignore its nature or thereafter its presence? Can he without loss avoid its contemplation? When we ask ourselves, toward the end of our life of efforts, what was it all about, we mean: what was there in the world, and what did we try to do in it? Action, of course, is what matters, but then action comes in all sorts of subtle varieties: not only crude overt action performed by the muscles, however significant these be, but also the action performed by the neural pathways, by the brain, of which, alas, we know yet so little. Feelings, thoughts — these, too, are actions but it is not enough merely to act, however incisive our actions. We must assure ourselves that what we are trying to do by our actions is worth doing and the best we could hope for, that it is, in fact, what we ought to do. And for this purpose we need to be critical of our own ontology.

To be aware of one's ontology is an enterprise of sufficient magnitude; but there is another task even more monumental, and that is its appraisal. Metaphysics is the criticism of ontology, and now our individual is to be placed in the position of a metaphysician. The piecemeal criticism of his ontology, that *is* his metaphysics. Some philosophers insist that metaphysics is the same as ontology, but that would be to claim that criticism and that which it criticizes are one and the same. Put two statements together, and you have implied a consistency which could be extended to others and so to a system. But take two apart, and devote your time to an analysis of them, to a logical and factual evaluation, and you have embarked upon a metaphysical enterprise. It should be noted that ontology is never used to mean metaphysics in this sense, but the term, metaphysics, has often been used to mean systematic ontology.

Logic looks at belief or knowledge from the point of view of its form; epistemology looks at it from the point of view of its evidence; metaphysics looks at it from the point of view of its content: how intense are its qualities, how well do they harmonize, how deeply do they penetrate, how much do they cover? Metaphysics assumes that there must be an ontology, with its own criteria of excellence, and then considers in the case of the ontology advanced by any individual how well these criteria have been met.

Obviously, no other ontology can be presumed to have passed the test, for otherwise it would have been adopted and the search considered ended. It is only when none is judged satisfactory that a new start is made.

Leaf through the early chapters of most contributions to philosophy. You will find there that the author has criticized his predecessors' views in order to make way for his own; for if the truth had already been discovered, what point would there be in embarking upon a voyage of discovery; or if truth were of such a nature that it had been compounded by all those great philosophers of the past who had found safety in what they asserted and jeopardy only in what they denied, what point would there be in undermining the achievements of the past so far as these carried with them any final claims? We are forever putting our beliefs together into a system as though we did not wish to have our actions defeat each other, but we are also under the necessity of retaining a continual criticism of the resultant system.

A quick glance at some examples taken from practical life should suffice to show what the average man has to gain from the philosophy with which we have been dealing so abstractly. Would a man who had examined his philosophy and made the appropriate changes still be willing to spend all his efforts in accumulating money and at the same time maintain a stout belief in a religion which held a low regard for the things of this world? Would it be possible for him to accept the tenets of any institution which had managed to perpetuate itself upon the insight that the world was coming to an end? Would he be eager to assign a large proportion of his income to the purchase of means of locomotion more powerful than he had any use for and more able to transport him than he had places he wished to go?

It is perhaps in the possession of an unique ontology and the ability to operate upon it by means of metaphysics that the source of each notable personality lies. Logically speaking, there are two sorts of relations between a man and his fellows: similarities and differences. The similarities are large and unimportant, and the differences small but crucial. At times, men take pride in being like all others, and at other times in being more of themselves. But there never was a time when a man did not want to count in the world for something, whether that something was his own or not. And to count means always to rely upon himself and his own attributes, to have the strength of his own virtues, to make his own decisions and, if possible, also to offer his own contributions. To this end, ontology and metaphysics can aid in a way no other study can.

Our next and last task is to suggest how ethics and aesthetics can be of assistance to the untrained individual. When we come to these so-called value studies, we are, so far as the individual is concerned, in the dimly lit area of intuitions about taste. There are of course systems of ethics and aesthetics; they are suspended halfway between systems of ontology and the relevant facts disclosed by individual experience, and must check with both. If the individual should start, as most of us do, with some notions of his own about what the good is, then he can find out through the study of ethics what his choice of values will involve him in. What ought he to hold in high esteem? What ought he to seek and therefore to do? These are the questions ethics endeavors to answer. Ethical systems to be effective must be society-wide, and they are so when they are openly acknowledged and established in codes of law. We shall return to this theme shortly. Suffice to say that the study of ethics can aid the individual in becoming aware of his deepest feelings and in showing him the extent of their involvement, often a way of assisting him in their correction. The individual will no doubt have unconsciously, or will acquire consciously, some conception of the ends toward which he strives, and these ends, so far as they are inclusive enough, are his highest goods. He will be enabled to strive toward them the more directly the more he is aware of what they are. The chances of a man arriving at where he is going are increased sharply when he knows exactly where it is that he wants to go.

At the same time, the means to be employed must not be left unconsidered from an ethical point of view. The use of immoral means to gain moral ends has been both praised and condemned— praised by those having moral ends which they thought would justify the means employed, and condemned by moralists on two grounds: first, that no amount of good can make an evil good and, second, that there is no guarantee that the employment of evil means will ensure the attainment of good ends, for if the evil means fail, only evil will have been achieved.

Ethics, we said earlier, is a study of the bond between wholes and so for the individual a study of the bond between himself and other persons and things. The name for this bond is goodness, and the right is a division of the good, since justice brings things into their own proper proportion, which is to say, into their own good. It must be remembered that every actual thing has bonds with every other actual thing. When that actual thing is an human individual, it is possible for him to recall that his thoughts, feelings and actions are in some ways trivial, for they are fleeting, infinitesimal things in a large and indifferent world, but in some ways crucial, since their ultimate influence is enormous and incalculable. The degree to which the crook of a finger upsets the atoms in the star Sirius has been over-advertised, but in the social world it is difficult to estimate the final reverberations of the effect of a single individual action. Therefore, for the individual to act without forethought, to deal without reflection on the portentious meaning of such dealing, is reckless in the extreme.

Ethics is a study of power in the realm of the good as it affects all people and things that share with us the world in which we live. So far as an individual's own values are concerned, he has to consider consistency and completeness here as much as he did in logic, only here completeness takes precedence, just as consistency took **precedence in logic. It would be a waste** if a man worked toward conflicting ends, for then all of his efforts would be self-defeating. And this happens. But equally important is the need to avoid having aims that fall short of what they should be; and this also and more often happens. Ethics is a study of the values which ought to be included in the aims toward the achievement of which we decide to devote our lives.

Aesthetics is a study of values of another though closely related sort: the bonds between parts within a whole, bonds which are felt and apprehended as the beautiful. When parts are perfectly fitted into the whole without shortcoming or remainder, they achieve a harmony and emit an effulgence or radiance as a result, so that the beautiful is fully described as the radiance of harmony. If the whole world were to be seen in such perfection, it would be the holy; but for this it would be necessary to stand outside. Art is the production of beauty. Every man has his own theory of art, just as he has his own theory of goodness; such theories range from the estimate that art is a worthless pretense (which is also a theory) to the estimate that art is the only thing worth living for. Systems of aesthetics are apt to be society-wide also, though not in the explicit and acknowledged way that ethics is.

For the productive artist there are of course special aesthetic considerations, but for the appreciator of art, which each of us is at some time or other, the human attitude toward art is a complex affair. For while there are personal gains from appreciating art, the chief end is not a selfish one. Art intensifies life in many ways; one of the best known of these is the education of the senses: for instance, we learn to hear by listening to music, to see from looking at painting, and to touch from handling sculpture. But for the most part our dedication is complete and without self-interest. The appreciation of art involves a feeling of superfluous caring. In art we love without wanting anything for ourselves, and care without having to be cared for in return. The effect of this on us, though not what we were after, is of the highest and most purifying sort.

For the individual, the value studies: ethics and aesthetics, enrich all experience. The study of the good deepens and extends all of his connections, and the study of the beautiful enlarges and rounds out his appreciation of the values of things. Neither goodness nor beauty asks the individual to consider himself; for the good he must 'do good' and for the beautiful he must admire: ethics calls for superfluous action and aesthetics for superfluous caring.

In general, perhaps we could say that philosophy magnifies existence for the individual. We live in as much time and space as we feel, know about or interact with. For certainly more affects us than we can affect: the forces with which we are in exchange are so much greater than we are, that our feeble influence is hardly felt; whereas we are the recipients of the effects of cosmic rays, of sunlight, and of those many bonds of society which lie beyond our power to reciprocate and which reach beyond our understanding.

It was just this understanding with which in the first place, we have claimed, philosophy is occupied. We are all affected by events which happen very far away in space, and this is true also of time: we are the inheritors of the entire past and of what has survived in it, and we are to some extent molded by all of society's plans and ambitions and unacknowledged efforts toward the future. If we are to be aware of the world in which we participate, we must know of the bonds which exist and of the forces which are at work; we must, in short, in addition to the special studies which give us the details, turn to philosophy both for a sense of the whole end and for the special tools by which to deal with our particular province of it.

V

How Can Philosophy Help Society?

No society can survive as a society for very long without some established principles and procedures. These usually occur in institutions: organizations of men and material which are usually, though not always, less than the whole of society. The principles and procedures embody philosophies; and sometimes this is done covertly but sometimes quite candidly. There is a philosophy of democracy, and it is embodied in the Declaration of Independence, the Constitution and the established procedures of our government, such as the system of political parties; but the democratic philosophy, derived as it is from Locke and Montesquieu, has never been explicitly established. There is a philosophy of communism openly adopted in the Soviet Union, and it is contained in the writings of Marx, Engels and Lenin.

Political systems are very wide affairs indeed when they are adopted by the governments of whole peoples; but there are other and more modest institutions which are important and which have their own established philosophies, whether open or assumed. Systems of law are underscored by jurisprudence; scientific principles and procedures are studied in the philosophy of science; and philosophies of art are advocated by particular movements in art; these are well-known examples.

Entire cultures, which extend beyond societies in the same way that societies extend beyond particular institutions, have their philosophies, though there philosophy has always been implicit and hidden. A single society embraces England, France and Italy in a way which brings out the resemblances between those countries; and if we were to examine closely the consistency between the various elements which they share, we should be well on our way to discovering the philosophy of the western society. Such efforts have been undertaken, though none perhaps with entire success: the task is too enormous, and we are too close. But from a distance it is possible to discern, however dimly, the philosophy of Asia, or of ancient Greece; distance lends perspective and therefore some grasp of the whole.

Now, the individual, in addition to having a private philosophy, shares in the philosophy of larger units. He shares for instance in the philosophy of some particular institution; he shares, to a lesser extent, in the philosophy of most of the institutions within his society; he shares in the philosophy of that society as a whole; and, lastly, he shares in the philosophy of the culture to which his society belongs. He is affected by all of these, and to some small or large degree he is capable of influencing them. It is necessary for him, therefore, to know their nature and if possible to be well-acquainted with them.

There are general reasons for this, for it is obvious that we would do well to know anything by which we were so greatly influenced and with which we are so profoundly involved. But there are in adition more special and at the same time more urgent reasons, particularly concerning the institutional philosophies. For institutions often adopt official philosophies, and official philosophy can be the death of philosophy. To adopt a philosophy officially means to have discovered the final truth; and it ends inquiry into the nature of truth by claiming the possession of it: we would never seek for that which we had thought we had found. Of course, a philosophy can be officially adopted without involving such perilous steps as the claiming of absoluteness and finality; but it usually does so, as past experience has shown.

What is the solution? There is an equally vicious danger lurking at the other end. Not to establish a philosophy means for a society chaos; without some degree of establishment there can be no stable society. Hence complete freedom in this sense allows for the absence of all order and security. It is not enough to have a personal philosophy, then, but there must be in addition a social philosophy which he shares and in whose benefits he participates. The answer ought to be some sort of tentative adoption of a philosophy which would provide for the requisite stability while allowing sufficiently for alternation and change to justify continued inquiry and progress.

What is called philosophy usually is what exists in a set of textbooks in libraries and what is taught by a set of teachers in classrooms who pass on to the next generations some of the abstract thoughts of the older tradition. Philosophy, in short, is a college course. But if this were all there were to it, no man with any intelligence or enterprise would wish to undertake it as a career. This, unfortunately, has only too often been the case. The topic is largely in the hands of men who believe just this, and so it has often had to be advocated by others whose professional competence lies in philosophy but whose living is somewhere else. The Greeks taught philosophy professionally, but since the Greeks many of the great philosophers of the western world have not been in universities. Averroes, Maimonides, Locke, Hume and Berkeley, to name but a few, were not university men, and in our own day Engels was not. Philosophers have most assuredly thrived in universities, and there are great names in this list too; Kant and Whitehead, for instance.

But whether the philosopher who is truly a philosopher, which is to say a philosophical explorer and not merely a teacher of philosophy, practices his profession inside or outside the university purview, the fact remains that traditions are advanced only by those who have no respect for them. In the end, every philosophy is found wanting. It could be the epitaph of every great systematic philosopher, "Here lies so-and-so, who failed where none succeeded." And yet the social world moves by means of such systems, and without them there could be no social world.

The task of philosophy is not confined to the past and hence not to the university. If cultures are applied philosophies, as we have been saying that they are, and no less so because silently and inadvertently dealt with, then the task of philosophy is to become a laboratory for the abstract examination of possible cultures. This is a tall order, but it does seem as though our field of philosophy must be considered in this connection. The cultural question of philosophy, then, becomes this: how ought we choose to affect ourselves?   What philosophical propositions can serve as cultural axioms whose consequences could be counted on to produce the kind of societies we think that there ought to be?

Admittedly, we lack the breadth of inclusiveness as yet to produce such effects. Even the experiment of trying to plan a society within the limits set by the axioms of a given culture has encountered serious difficulties. Soviet Russia has laid down for itself a goal which is well within the limits of the axioms of the western culture: a society organized unilaterally and existing for the maximization of the institution of applied science. Applied science, and Marx's theories, too, for that matter, were western cultural inventions, and whatever society utilizes them on a grand scale must come within the purview of established events of the type of the western culture.

Philosophy has many such tricks up its sleeve, and its task as a theoretical laboratory is to explore the abstract possibilities, for speculation is a sort of looking ahead at what practice could be. Its vitality as an intellectual discipline is dependent upon the scope with which it is envisaged as well as the incisiveness of its own professional tools. But we must acquire some proficiency in the use of the latter before we are ready to undertake the tasks of the former.

Thus the philosophical enterprise finds its justification in the necessity for individuals to explore their own beliefs and for societies to provide stability and progress. It could hardly be more important. And if to the plain man working for his living it does not seem to be of day to day relevance, that is because the prominence of importunateness has driven away the larger background of importance. Hunger always presses on us harder than curiosity, even though it was curiosity which led to the civilized improvements in the manner in which we satisfy our hunger. The cumulative process is never noticed yet takes its toll; we do not get much older in a day, but after many such days we die. We are no less the victim of the massing of effects for not noticing single ones as they occur. The man of breadth, then, will attend to philosophy, when to do so means living more fully the small life that we have.

VI

A GLANCE BACKWARD

There is no benefit that can be achieved without the cost of some effort. Learning often is, at least to those who have grown accustomed to it, a delight, but to those with no practice it offers all the awkwardness and the pain to be found in the unused muscles of the beginner. The concentration upon abstract ideas so intimately curious is not easy for those who have not habitually done it, and thus the prospect is a repelling one. It seems very much easier to do something else with your time. Philosophy in this respect is no different from ice-skating or the playing of contract bridge. It is only the rewards that philosophy offers, which are greater than others, that must be the inducement; for each of us has, if he could only become aware of it, a need to know.

To become keenly conscious of the existence and the strength of the need to know, is to begin the long task of attempting to satisfy it. It will be found to be like the other needs, like hunger, for instance, which can be satisfied only for a time, and which arises again in all its demands if left alone for very long. We cannot hope to know what there is to be known; very little knowledge of anything is yet the possession of the entire human race. But we can make a start in learning what is known or in prospecting for ourselves beyond the frontiers of knowledge. Learning is not a process with a marked limit; it is a way of life, and those who live in this way live more intensely and have the experience of a fuller existence.

Those who have followed the suggestions in this essay, and have tried to put themselves into the discipline that devotes a small part of every day to the practice of philosophy, first learning something of the history of philosophy, then studying the nature of the separate philosophical subdivisions, and finally attempting to discover and perhaps to revise what they themselves believe to be true at the profoundest level of belief, will be astonished after a time, a year or two, say, to discover upon looking backward how far they have come. And they will be astonished, too, to discover how much the plain events of everyday life have acquired a meaning that they did not have before, and life an enjoyment they did not dream it could possess.

They began, because we all do, by holding a philosophy consisting of prejudices acquired at random through the emotions and defended by the use, which may at times even be skilful, of fallacies. But now they can be satisfied with nothing less than reasons, reasons which derive from facts carefully sifted for their value as evidence, or from other reasons which had already been well established in the same fashion by derivation from the incontrovertible principles of logic. And there will be from this a qualitative change which imperceptibly grew upon them, and will raise them in fullness to the. stature of aware beings in the highest degree of what it means to be human and themselves.

PHILOSOPHY AND THE CATEGORIES OF EXPERIENCE

Harold N. Lee

PHILOSOPHY, in its broadest and least technical sense, is a world-view—a way of looking at the world; and the way that one looks at the world depends on the categories by means of which he approaches, classifies and interprets it. For example, a religious philosophy classifies what it finds under religious categories. A scientific philosophy holds that various sorts of natural phenomena, their processes, relationships, and the regularities to be discovered therein are fundamental. A materialistic philosophy seeks to classify all experience as the appearances and interactions of matter; and so on.

In a more technical sense, philosophy is a systematic attempt to understand experience, and understanding consists in the grasp of logical relations. Logical relations subsist not so much between things as between classes, or kinds, of things; and the fundamental, or master, classifications are categories. Hence, here again, the categorizing activity of the mind is seen to be fundamental to the philosophic enterprise.

The thesis of the present essay is that the philosophic enterprise, in its most technical sense, is the formulation, development and criticism of the categories by means of which the understanding of experience is made possible. This definition indicates the purpose, the methods and the result of the philosophic enterprise. Understanding is the purpose, and it is attained by a critical development of categories. Understanding is a matter of generalization, of classification, of division, of distinction, of definition. It depends upon the grasp of relations, and clear cut relations are not to be found without clear cut categories. Definitions are not of things, but of classes and relations. When the raw materials of experience are formulated, distinctions are found or introduced; and when this process is carried to a clarified conclusion, the distinctions crystallize into definitions. The most fundamental definitions—those of greatest generality—are definitions of categories. Categories are classifications of widest scope; they are complexly interrelated; and to the degree that we can subsume experience under categories which enable us to grasp its relationships, to that degree we have understood it. A system of philosophy is a categorial scheme of great complexity in which the categories are logically ordered and are relatively adequate to cover experience.

Whenever, in any field of human knowledge, the widest categories of that field are criticised, reformulated or altered, the philosophy of that field is under consideration. For example, the concept of number is one of the fundamental categories of mathematics, and when one investigates the nature of number and thus of the mathematical processes involving number, he is engaged in the pursuit of the philosophy of mathematics. Similarly, the category of cause and effect is fundamental in physics, and the philosophy of physics has changed when the mechanical concept of the relation between cause and effect gives way to the statistical concept. Again, St. Augustine's philosophy of history differs from that of Marx in that to St. Augustine all history is the outcome of will, or decision, that is shown by the inner light of consciousness to be determinate of events. Here the categories of will and the inner light of consciousness are dominant. To Marx, on the other hand, all events seemed to be the outcome of what he called the inexorable law of material processes. Here 'matter', 'process' and 'law' name the basic categories.

Thus, the task of philosophy is to understand the widest interrelationships of experience and to formulate the tools whereby such understanding can be attained. The tools are the categories. It is not the business of philosophy to find out what things are in the world; this is the task of the special sciences. In performing this task, the special sciences operate with categories already established: for example, 'thing' and 'the world'. It is not even the task of philosophy to find out what things are real. Everything is real in some sense. It is the task of philosophy to find out in what sense or senses. This is where the categories come in. Whatever is in question is real in some category or other. If we can get the categorial scheme straightened out, and subsume that which is in question under the appropriate category, it is real in that category. For example, an experience that can not be subsumed under the category 'physical object' may nevertheless be subsumable under one of the categories 'dream' or 'illusion' or 'hallucination'. There are real dreams, real illusions and real hallucinations. If one pretends to have an hallucination which he does not have, there is nevertheless a real pretension or prevarication. 'Phantom' and 'delusion' are other categories that take care of the reality of what can not be subsumed under categories such as 'actual situation'.

The task of philosophy is not exhausted by finding the appropriate category, however. The complex interrelationships between categories must be established. In terms of the illustration, the relation between 'dream', 'hallucination', 'actuality' must be found. Science tends to assume that the category 'physical object' or 'physical process' is more fundamental than 'dream' or 'illusion', and to explain the latter in terms of the former. The critical investigation of this assumption is a truly philosophic task as it is fundamentally concerned with the criticism of categorial schemes. It is the business of philosophy to coordinate all categorial schemes so that we can understand how all experiences go together and how the special sciences contribute to total knowledge.

Cognition takes place in terms of concepts. Concepts are general; that is, are of classes or kinds or of relations. Classes can not be defined without relations and general relations can not be defined without classes. The widest and most inclusive classes are categories. The difference between concepts and categories is one of relative degree, not of kind. Categories are concepts of wide scope. They are those concepts that are most fundamental to understanding. Because categories are a subclass of concepts, it is not the task of philosophy to construct and present a list of *the* categories. Alternate categories are possible, and philosophy is the criticism and elaboration of whatever list is presented. A philosopher may develop his own list, but in doing so, it is incumbent upon him to remember that he is not the only philosopher.

The clear comprehension of a category requires that we know how to apply it; that is, what experiences can be subsumed under it. Consequently, the study of the categories requires constant testing and application, and this is a practical matter. The application is not made for the sake of practical results, however, but only for the sake of the development and criticism of the categories. For example; it is not the business of ethics, in so far as ethics is a branch of philosophy, to make individuals act morally; nor is it the business of social or political philosophy to make the world good. Rather, it is the business of these disciplines to understand clearly and completely what it means for an individual or the world to be good. It is not the business of epistemology to produce knowledge about anything but the nature of knowledge—knowledge of the categories that enable us to understand the knowledge process. It is not the business of logic to produce absolutely consistent thinking on all subjects; but it is its business to formulate and articulate the relations that would be involved in absolute consistency. It is not the business of aesthetics to stimulate the production of art or even the appreciation of beauty; nor of the philosophy of religion to determine the true religion and infuse men with its spirit. Philosophy is the formulation and criticism of categorial schemes whereby these areas of experience can be understood.

II

The present theory of the nature of the philosophic enterprise can be supported only by presenting a categorial scheme of categorial schemes; that is, a categorial scheme which itself makes intelligible the formulation, development, criticism and application of categories within experience. No one is born a philosopher, but as soon as he learns to speak, he adopts and uses categories which are embodied in language. He may, as he grows older, begin logically to criticise and extend these categories, and as soon as he does so, he is embarking on the philosophic enterprise. The enterprise must start somewhere, and it can not start (in point of time) from fully developed categories, for to develop such is its goal. As Santayana pointed out, one can start only from where he is,[1] and the beginning philosopher is an adult human being who lives in a world of ordinary perceptual objects. The objects are of kinds or classes already blocked out for him in the language he socially inherits. This is the world of common sense, and if common sense categories rendered all experience intelligible, there would be no philosophy beyond common sense.

There are several flaws in the common sense picture, however. There is more in experience than can be subsumed under common sense categories: they are too limited and parochial. Again, common sense categories do not permit the degree of control and pre-

---

1 See Chapter I of *Scepticism and Animal Faith*, N.Y., 1923.

diction of experience possible with the use of more elaborate categorial schemes. In addition, common sense categories are not always consistent with each other. Finally, we often make common sense errors: the categories break down and do not render experience intelligible.

If the categories of common sense are not adequate to the task cut out for them, the thing to do is to take them apart and see what is the matter — see why and in what respect they are inadequate. The technical word for this is 'analysis', and analysis is the most important tool of criticism.

I wish explicitly to avoid several assumptions often uncritically made concerning philosophical analysis. I do not assume that it is the purpose of analysis to find separate or separable parts out of which whatever is analyzed is made. It is true that I am using the term 'analyzing' to mean only 'taking apart', but the taking apart is in thought: the parts do not have to be exhibited separately in order to insure the legitimacy of the analysis. If we are able to formulate concepts of parts that can be shown to vary independently of each other, the analysis is justified. The basic method of philosophy is logical analysis; that is, the analysis of concepts.

A second assumption often uncritically made about the process of analysis is that it gets down finally to one set of ultimate parts: the correct analysis finds the one set and all others are wrong. This assumption seems to me not only to be unwarranted but to fly directly into the face of the evidence. Alternate ways of taking a given complex apart in thought can always be found. Any way of getting it apart is "correct." Some ways may be better in accord with some purpose or may be more successful and some less, but the only "wrongness" in this matter is indicated by failure.

A third assumption, often combined or even confused with the second, is that the ultimate parts found by analysis are simple and discrete; simple in the sense that they are not subject to further analysis because they contain no parts; discrete in the sense that each is an ultimate unit not continuous with any other. The view presented in the present essay holds that no evidence can be adduced that an absolute simple has been reached; all that can be said is that further analysis has not been successful. As for the question of discreteness, Zeno long ago pointed out that if one starts with discrete parts, he can never get a continuum. Dedekind showed, however, that if one starts with a continuum, he can get discrete parts. Consequently, I avoid the uncritical assumption of ultimate, simple, discrete units as the outcome or goal of analysis.

It has sometimes been assumed by reputable philosophers that analysis, completely carried out, yields the "ultimate constituents of the world."[2] Such an assumption rests on an inadequate criticism of common sense categories, for it assumes uncritically both 'the world' and 'ultimate constituent'. It may be all right to assume 'the world' as a starting point, but if we do, we ought not to say anything ultimate about it — at least not in the same breath — for 'ultimate' is a category not properly applied to a starting point. There may be ultimate parts or constituents that are discrete and simple, but it is not the assumption of the present essay that the goal of correct analysis is to be stated in these terms. It is the strong suspicion of the present writer that the combination of categories expressed in the phrase "the ultimate constituents of the world" has no application. The purpose of analysis is to break whatever we are investigating up into parts so that recurrences, resemblances, patterns, form can be discerned—so that abstraction can take place. There is no understanding without abstracting.

When we think of one of the parts found by analysis by itself, without reference to the context in which we have found it, we have abstracted. This is all that 'abstraction' originally means— taking something out of the context were we found it. 'Abstraction' names a process. The significant purpose of analyzing is to be able to abstract; and the significant purpose of abstracting is to be able to generalize. When we find that what we have abstracted from one context can also be abstracted from others, we have generalized. A generalization is what is common to more than one instance.[3] Generalization yields class concepts, and the class concepts of widest scope are the categories by means of which we approach the interpretation of experience. Common sense categories are constellations of generalizations arising from ordinary un- criticised and unanalysed perception; and without them, the per- ception would not be exactly what it is.

It is no more to be supposed that the products of abstraction can exist by themselves than that the parts found by conceptual

2 The expression quoted is from Russell's "Philosophy of Logical Atomism"; originally published in the *Monist* as a series of articles, Vol. 28 (1918) and Vol. 29 (1919); republished by the Department of Philosophy of the University of Minnesota. See Lec- ture VIII (p. 58ff of the republication).

3 For a more detailed theory of the relation between analysis, abstraction and generalization, see the author's essay "An Epistemological Analysis of Induction" in *Tulane Studies in Philosophy*, Vol. II (1953).

analysis can be separately exhibited. Abstraction does not mean actual separation. Many philosophers have inveighed against abstraction, but without abstracting, there is no generalization, and without generalization, there are no concepts. What the philosophers have warrant to inveigh against is *abstractions* when the word is used to name the hypostatized products of the conceptual process of abstracting. I am not sure that there are any such *things* as abstractions, but without *abstracting*, there is no cognition, because cognition involves class concepts, and concepts depend on being able to pick out (abstract) what is common to many specific instances. The error to be guarded against is what Whitehead called "the fallacy of misplaced concreteness."[4] This fallacy itself is an error of categorization. Thus, in point of temporal origin, the philosophic enterprise begins with the criticism of common sense categories, but these are not necessarily logically fundamental. The categories that are at the bottom of understanding must be logically fundamental, and these can be found only by the aid of analysis.

III

Let us proceed to the analysis of the common sense world from which the philosophic enterprise must start not logically but in point of time. 'Experience' may be used in two very different senses: 1) to indicate the formulated experience of ordinary perceptual objects and events; or 2) to indicate the raw materials out of which the formulated experience (ordinary perception) is made. The first is already present when one starts philosophizing; the second is found only by analysis. The double use of the term is unfortunate, but may be justified because although the two senses can be distinguished, there is no absolute dividing line or line of separation between the two kinds of experience indicated. Unless the distinction is made, however, confusion or equivocation results. It can be made and clearly held in mind in spite of the fact that an adult human being is rarely (or perhaps never) aware of the raw materials as raw — that is, as unformulated and uncategorized. He nevertheless finds different degrees of formulation in his awareness. Furthermore, it is difficult to see what it would mean to talk of formulated experience unless there were something there to formulate.

The raw materials of experience are roughly of the nature of that of which we are aware in sensation, but it is obvious that I

---

4 See A. N. Whitehead, *Process and Reality*, London, 1929, p. 11. See also *Science and the Modern World*, N.Y., 1926, p. 85.

have already categorized and interpreted them in identifying them. One can not do otherwise. To say anything whatever about the raw materials is to subsume them under concepts. It is not an accident that 'communicate' and 'common' are cognate terms. Only concepts are common in knowledge.

It may be that a new-born infant is aware of the raw materials of perception as raw, but a new-born infant does not philosophize. Nevertheless, the new-born infant must be aware of something, and this same sort of something must be present in all perception whether or not as adults we can find it only by analysis and are not aware of it separately. Otherwise, there would be no feeling of living in a world other than ourselves. The only reason that we are not all solipsists is that there is something in our experience that seems to come from outside; but when I analyze the 'we' of this statement, it turns out to be a way of saying that I find that something *does* impinge upon my awareness from outside. I have the suspicion that anyone who denies that there is some ultimately given content in perception is a solipsist; but this is only to confess, that I am not.[5] Something is ultimately given — other things are really other; and in this statement 'really other' means correct subsumption under the category 'other'.

The expression 'given in experience' is subject to the same ambiguity as was remarked three paragraphs back for the word 'experience'. I am using the expression to refer to the so-called raw materials, that is, the content or ultimate ground for my holding that there is anything subsumable under the category 'other'. If it is really other, I receive it from without; but I am aware of it only in so far as I assimilate and interpret it. It is given in the sense that I receive it but not in the sense that I am originally aware of it. It can be found by analysis, however; thus, it can be called 'the analytically given'; but for brevity, I shall drop the qualifying adjective where no misunderstanding can arise from its omission.

An adult is immediately aware of the formulated experience of the objects of ordinary perception, and these are the data from which we start philosophizing. We do not start philosophizing from data that are analytically given. The analysis itself is a philosophic activity. Thus, there is warrant for calling ordinary perception 'given'. Nevertheless, I do not follow that usage, but by 'the given

---

5 If there is anyone who is not a solipsist, then I am not a solipsist; for if the 'anyone' is I, then I am not a solipsist; and if the 'anyone' is not I, then I have posited an other.

in experience' refer to the analytically given. The immediacy of ordinary perception is immediacy only in time: ordinary perception is temporally but not analytically immediate. Ordinarily, the interpretation of the analytically given is not separated from an awareness of that given by a lapse of time. Except in rare instances, an adult is aware of what is given only as it is interpreted — only as it is taken up into a ready-made conceptual framework; and it is only as it fits into some such framework that orderly experience, or perception of fact, results. The raw material of perception — that which is seen by analysis to be ultimately given — may be called 'perceptual intuition'; and the formulated perception found in ordinary experience may be called 'perception of fact'.[6] The use of some such terms as these to distinguish between the two moments of so-called immediate experience will help avoid equivocation or confusion.

It is not assumed that there is any absolute dividing line between perceptual intuition and perception of fact. They are terms of analysis and are not to be looked for separately in experience. Nevertheless, that the analysis is reasonable is shown not only by the evidence already adduced but also by the consideration of perceptual error. For perceptual intuition, error is a meaningless concept; for perception of fact, it has import. We interpret some data of perceptual intuition to be a deer in the woods, but we sometimes have occasion later to change the interpretation (with some chagrin) to be a rotting stump and log upon which the early morning light fell in a strange way. Or we see a man standing in front of an automobile on a dark street, but more careful interpretation shows an unusual pattern of moonlight through the trees falling on the automobile. In all such cases of perceptual error, and in perceptual illusions and hallucinations, something is presented to awareness but is misinterpreted. It would involve a contradiction in terms to say that what is ultimately given is erroneous. If it is given in any ultimate sense, it is there — it is only what it is. Error is relative to knowledge and arises as does knowledge when what is given is categorized. If it is well categorized, the product is knowledge. If it is badly categorized, the product is error.[7]

---

6 These terms are taken from my earlier work on aesthetics, *Perception and Aesthetic Value*, N. Y., 1938. See Ch. III, Sec. 4; but I would not, outside the context of aesthetics, stress awareness of perceptual intuition. The artist and the appreciator can train himself to such awareness, but it usually takes training.

7 As C. I. Lewis points out, the given in perception is instanced by what we see when we see a deer when there is no deer. See *Analysis of Knowledge and Valuation*, LaSalle, Ill., 1947, p. 183. Perhaps we later decide that what we saw was a rotting stump, but that is an interpretation just as the deer was. The given is whatever it is that is being interpreted.

It would be a mistake, however, to identify the given with sensation or sense data. It may be that the given is roughly of the nature of that of which we are aware in sensation, but they are not identical. Every consideration that makes it reasonable to hold that sense data are given applies also to the content of sense imagery, dreams, hallucinations, illusions, and so on. There is no ultimate difference in kind between the raw material of actual sensations on the one hand and the raw material of dreams or hallucinations on the other. 'Sense data' refers to the actual stimulation of the sense organs, and the given does not come already blocked out into parts that are actual and other parts that are not actual on the face of it. The distinction is categorial. The given can be subsumed under the categories 'actual sensation' or 'veridical perception' only by virtue of relations of coherence and orderly sequence that go beyond the given—that are products of the analyzing and abstracting activities of the mind. What we mean by 'orderly experience' is the successful application of categories to the given. Dreams and hallucinations become parts of orderly experience as soon as the categories and their application are straightened out.

Even if we are not aware of the given uncategorized, this does not imply that it is unknowable. It is eminently knowable. We subsume the given under categories, and this is knowledge. Originally, there is no other source of knowledge. The subsumption under categories is hypothetical, and this is the sense in which all knowledge is hypothetical. Epistemology makes the hypothesis that the data of perceptual intuition are subsumable under categories, and this hypothesis is verified to the degree that experience is thus rendered orderly and intelligible.

Another hypothesis that it is reasonable to make about the data of perceptual intuition is that they are a continuum. No absolute dividing lines or lines of separation are given. The given is not made up of discrete parts. If we were postulating that the given is made up of discrete parts, we could not consistently have said that analysis and abstraction do not necessarily arrive at discrete units. If the given were made up of discrete parts, then the only legitimate analysis would be that which reached or reached toward those parts.

To say that the given is a continuum does not imply, however, that it is undifferentiated and characterless. It does imply that all differences grade into each other without interruption, separation or absolute dividing lines. The spectrum of colors is a continuum, but that does not make it an undifferentiated sameness. Red is still different from yellow in spite of the fact that no absolute dividing line is to be found — that at no place can it be said "here red stops and yellow begins." The continuum can be cut at any place desired, and its character will be different at each place it is cut.[8] The difference between a continuum and a discrete series is that in a continuum, the exact places of the cuts are not preestablished. The cuts can be anywhere.[9]

Categories, classes and kinds are generalizations of cuts made in the continuum of the given—they are the result of the logical activity of the mind. Both the intuitive and the logical processes of the mind are involved in cogition; but the mind is not here conceived as a substance, it is rather conceived as a process or an activity of reacting to the environment. The mind is an instrument of adjustment and adaptation to the environment. It operates to produce knowledge, and the tools of knowledge are concepts. The continuum of the given is subsumed under concepts which represent cuts in it; and the concepts are related to each other by the logical relations of consistency, inclusion and exclusion. Concepts are of the mind, and the mind in taking hold of the continuum subsumes it under concepts. In so far as the subsumption is adequate and clear, and as the concepts are consistently interrelated, experience is understood. The subsumption is not in any way arbitrary, however, for whether or not the continuum can be subsumed under a special concept depends on the nature of the continuum.

The present theory is broadly Kantian, but without agreeing with the details of Kant's theory. It holds that Kant's fundamental mistakes lay in two places: in supposing a "thing in itself"; and in supposing a rigid system of categories. Kant held that the categories are determined by the structure of the mind. He was working within the Cartesian framework and conceived of the mind as substance; hence, it had attributes and a structure. The structure was permanent and universal; otherwise the substance would not be enduring. If, however, we conceive mind to be a process or activity, there is no call for a permanent structure, and hence no

---

8 I am using the analogy of the Dedekind cut, but that term was evidently used by Dedekind because of the more obvious metaphor.

9 One of the fundamental properties of a continuum (the Dedekind property) is that it can be divided into two parts such that any element belongs either to one part or the other exclusively, and this division can be made anywhere. See E. V. Huntington, *The Continuum and Other Types of Serial Order*, Cambridge, Mass., 1921, Ch. V. Postulate C1 on p. 44 states the Dedekind property.

rigid categories.[10] If the categories are not fixed, either in the nature of things or in the structure of the mind, there is no call for a *list of categories*. Some categories are always developed when the cognizing activity of the mind is at work, but *what* categories depend on factors many of which are incidental to the nature of the mind. If the mind is a way of reacting to environment, then the effective source of the development of categories is the need of "finding one's way around" in the environment.

IV

In the history of the race, the development of categorial schemes was doubtless the result of long processes of trial and error in dealing with the environment, and was bound up with the emergence of language. Common nouns represent concepts. Concepts arise from the cuts made in the continuum in the process of reacting to it. The cuts, or rather, the manner of cutting, thus become crystallized in language and are handed on to succeeding generations. No one person ever constructs his own conceptual schemes. In learning language, he adopts the conceptual schemes of his forebears. He may (and perhaps usually does) modify what he receives, but his concepts are rarely his own inventions. Hence, the cuts that we habitually make in the continuum are not to be called subjective.

Neither is the continuum to be though of as subjective. So to think of it would involve a confusion of categories. The category 'continuum' as used in the present essay has been closely connected with 'otherness' and 'what there is' or 'what is there'. The difference between subject and object is a distinction made *within* the continuum. Thus, 'subject' and 'object' name sub-categories. In so far as the difference between subject and object can be congnitively grasped, it is grasped by definitions. The continuum does not depend on the definitions of subject and object, however; instead it is the ground of them.

All concepts are grasped by definition, and definition is limitation. A definition puts boundaries, and if the definition is precise, the boundaries are exact and clear. Concepts, thus, are essentially discrete. Concepts arise from cuts made in the continuum, and the cuts are the bases of discrete divisions. The essential character-

---

10 See the author's essay "The Rigidity of Kant's Categories" in *Tulane Studies in Philosophy*, Vol. III (1954).

istic of the continuum, however, is that the division can be made *anywhere*. Thus, alternate sets of concepts may deal with the same continuum, but logic demands that the concepts be constructed in an orderly pattern. The concepts must be interrelated and organized by means of relations of inclusion, exclusion, equivalence or diversity of intension. Only if they are so related will they perform their task of enabling us to grasp the continuum.

The basis for the orderly relation of concepts is to be found in the continuum itself. It has been pointed out above that a continuum is not necessarily an undifferentiated sameness. There may be differences, and these differences may themselves fall into a pattern. Take a simple wave form for an illustration. In any simple wave, such as a water-wave, there are not only differences, there are repetitions; but as these repetitions are in a continuum, no fixed boundaries are pre-established. For example, a wave may be regarded as a series of successive convexities; but it may equally well be regarded as a series of successive concavities; or as a series of alternating convexities and concavities; and so on. To the question "But what are waves, really, successive convexities, successive concavities, or what not"; the answer must be given "Any one of them as much as any other." If the continuum is analyzed in one way, they are the one; if it is analyzed in another way, they are another. No single analysis (in this illustration) is either more or less true to the continuum than another.

The continuum of intuited data is, of course, far more complex than the illustration of a simple water wave would indicate. To carry on the illustration, multitudes of waves may move through the same medium, and in doing so they form nodes. These nodes may be very pronounced, but they are not discrete units. They mark the repetitions and patterning of the continuum. When concepts are based on cuts made at the nodes, they can be said to "cut reality at the joints." When these concepts are subjected to criticism and adjustment, they comprize categorial schemes for understanding the continuum.

The concepts are discrete but the nodes are not discrete; thus, there is not one and only one way of cutting reality at the joints. Other categorial schemes based on other analyses will find other joints; and as long as the different concepts involved in the different categorial schemes are consistent and logically related, it can not be said that one scheme is true and all the others false. *Any* way of cutting the continuum yields only a selection of what is there no matter how numerous or frequent the cuts; and the character of the continuum may seem to be quite different according to how it is cut, for the selection is different. Some selections are better than others in enabling us to understand and control the continuum, and the faith in logic is the assumption that concepts which are orderly and are precisely interrelated will be better. This faith is largely justified in the outcome of human affairs, especially in such fields of experience as natural science, which has been eminently successful in evolving highly logical conceptual schemes that can be applied to prediction and control over environment. There is evidence in other fields also that understanding contributes to control over environment. Natural science has discovered that the nodes marking the repetitions and patterns of widest scope are not easy to find, but when it has found them and has based its categorial schemes upon them, it has been eminently successful.

Ordinary empirical concepts — the concepts involved in ordinary perception of fact — are based on the obvious or most easily apprehended nodes. The apprehension of nodes in the continuum is due to noticing repetitions or patterns in it or by being struck by "thickenings" or "thinnings" in it. As has been pointed out, however, the development of concepts involves the processes of analysis, abstraction and generalization. The concepts of ordinary empirical perception of fact, then are made by analysis, abstraction and generalization of cuts made at the obvious nodes; but when we have become familiar with the development and manipulation of concepts, we can carry the processes of abstration and generalization farther and farther until the concepts of pure mathematics and logic are produced. It is to be supposed that the concepts of natural science are constructed on the basis of observed nodes in the continuum, and that these are found, not made; but the concepts of pure mathematics and logic are not to be construed as necessarily depending upon or referring to anything in the continuum. The existence of branches of mathematics and logic that have no known application to experience is evidence of this. We have learned so well the process of constructing concepts that we can finally free them of the material out of which they were originally constructed.

All concepts, because of their manner of construction, are hypothetical in their application to the continuum. That is to say, all analytic and theoretic knowledge of the continuum is hypothetical;

but the continuum itself, in all its infinite diversity, repetitions and patterns is not hypothetical. In its own nature, it is neither subjective nor objective. Instead, it gives rise to the distinction between subjective and objective — the ground of the distinction is to be found only within it. In its relation to knowledge, however, the continuum is objective; it is conceived to be the only object of knowledge in the sense that it contains all objects of knowledge. The flux of intuited data is that from which all concepts, categories and categorial schemes arise and also is that to which they all refer except in the case of the concepts of pure mathematics and logic.

V

It may be well to digress for a moment to illustrate the thesis of the present essay by reference to episodes taken from the history of philosophy, for the history of philosophy itself may be regarded as the history of the formulation, development and criticism of categories. In the Greek beginnings, the vague concept 'arche' gradually became the category 'substance'. Then 'process' or 'process of change' was added as a result of Anaximander's suggestion of "separating out," and more specifically by Anaximines' "condensation and rarifaction." Heraclitus and Parmenides added 'the flux', 'appearance', 'permanence' and 'illusion'. Next came 'combination' or 'interaction', and the suggestion of a category we would now call 'emergence', but it was not developed. Democritus finally added the category 'the void' and had the scheme for systematizing all the early philosophy.

The early movement had neglected categories of knowledge and of value. The Sophistic period and Socrates developed these. The philosophies of Plato and of Aristotle clarified and systematized the humanistic categories and, further, attempted to co-ordinate them with the cosmological categories of the pre-Socratics.

Christian philosophy introduced the category of historical process. Augustine added 'the will', and from this was developed the categorial opposition of 'objective' and 'subjective'. This opposition was hypostatized by Descartes into the substances mind and matter, and what Whitehead called "the bifurcation of nature" took place. In the light of the present essay, the bifurcation must be viewed as due to a categorial confusion. The discreteness of the categories by means of which reality is grasped was imported into the reality: and Descartes conceived of two separate realities which are not continuous.

Much of the philosophic ingenuity of the next two centuries was devoted to solving the problems that stemmed from this confusion. Spinoza saw that whatever it is that is categorized in human knowledge is a continuity, but 'substance' remained the master category for him, and mind and matter became sub-categories of substance. Since they were discrete, they could not be related by continuous causation, and 'parallelism' was his fundamental category of relatedness.

Leibniz broke substance up into an infinite number of discrete units — the monads. Continuity and relatedness must be accounted for, however, and he accounts for them by the universal parallelism of the pre-established harmony based on his doctrine of internal relations: monads can not causally interact but they do mirror each other. 'Mind' and 'matter' are no longer of fundamental importance, but give place to 'activity' and 'passivity'. All monads are relative degrees of activity and passivity; thus, substance *is* process and relation. Herein lies another categorial confusion. Leibniz did not need 'substance', but he could not get away from it because it was firmly established in the philosophic tradition of his age.

It follows from the doctrine of internal relations that if one monad were different from what it is, all other monads would have to be different. Leibniz took this into account by developing a category of cosmic possibility worked out in his theory of compossible universes of which the actual one is the best. The categorial scheme sketched in the present essay presents alternate systems of compossibility much like those of Leibniz except that the notion of discrete units of substance — the monads — is given up and the categorial confusion between 'substance', 'process' and 'relation' is thus avoided. Instead of monads, the present categorial scheme has nodes in a continuum. If cuts are made so that the resultant concepts are logically consistent and systematic, one compossibility of understanding results. If cuts are made at other places so that a different set of consistent and systematic concepts is evolved, a different compossibility results. Both sets of cuts can be made with regard to nodes, for nodes in a continuum are not discrete. Instead of different cosmic compossibilities of which only one is actual, we have different compossibilities of understanding within the same continuum. 'Possibility' is a logical category, not a cosmic one; and compossibilities are alternate structures of understanding, not of actuality. Different compossibilities can co-exist and do so. Every different system of philosophy in the history of philosophy (i.e. every set of logically interrelated categories) is a compossibility, though none is logically perfect.

Locke and Hume used the method of analysis in a way very different from that in which it was used by Descartes, Spinoza and Leibniz. Locke and Hume applied analysis to experience, and attempted to find the ultimate materials out of which knowledge is constructed. The materials they found, however, were discrete units, and as Kant pointed out in adverse criticism, were conceived to come to the mind ready made. Thus, to Locke and Hume, the materials of knowledge are already categorized. With this analysis, it is no wonder that Hume could find no basis in the materials of experience for causation. The conception of discrete units which are related by continuous causation involves a logical inconsistency.

Hegel's influence on the subsequent history of philosophy was profound and complex. One of the important aspects of his thought was the development of a categorial distinction between reality and appearance. Many of his followers assumed that it is one of the tasks of philosophy, if not the most important one, to *distinguish* between reality and appearance. This distinction can be made, however, only if 'reality' and 'appearance' are both names of categories. According to the view of the present essay, to think of them both as categories involves confusion. 'Reality', instead of being a category either parallel to or inclusive of 'appearance', is not a category, at least not in the same sense as the other categories. If it is to be called a category, it is the wholly ubiquitous one — its extension is universal and its intension is null. 'Reality' names that which either is or can be categorized. An appearance is any categorization of reality. Appearance is a proper part of reality: all appearances are reality, but some reality is not appearance — namely, the reality which is not yet known. Reality, being a continuum, is infinite, and as infinite, any proper part of it is equally real with the whole. Reality is subject to rational understanding. It is the name of whatever is so subject. Rational understanding is achieved by means of subsumption under categories, and to talk of a reality which is either not amenable to the categorial process or is beyond it is to deny the cogency of the process. It is not apparent that the term 'ultimate reality' has any special meaning. If it means 'uncategorized reality', then it is what is at present unknown but nevertheless subject to knowledge; it is not unknowable unless it is impossible to subsume it under any category. If and when it becomes known, it is subsumed under categories, and thus, becomes appearance. In this case, as soon as we know ultimate reality, it becomes not ultimate. This is categorial nonsense.

In recent philosophy, the system that takes continuity most seriously is Bergson's. Although Bergson finds analysis and categorization necessary to action, he holds that they yield only practical (or, in its highest development, scientific) knowledge. Aside from the point that his view misreads the nature of science, it is unsatisfactory in that it misreads the nature of knowledge by relegating the categorial process to a special field. Bergson postulates a radical difference between practical knowledge and philosophic knowledge. He holds that intuition is knowledge. The view of the present essay is that intuition is not knowledge but yields the data out of which knowledge is made. Intuitive data yield knowledge when they are subsumed under categories; and this is true in the case of philosophic knowledge for Bergson as well as in the case of practical knowledge: the categories 'intellect' and 'instinct' are fundamental in his theory of knowledge.

If only intuition without categorization yields philosophic knowledge, mysticism is the only philosophy. On the other hand, if categorization is essential to the knowledge process, mysticism is not so much to be called philosophy as a surrender of philosophy. It is not the task of philosophy to distinguish between one kind of knowledge which is philosophical and other kinds which are not. Knowledge is continuous in all its kinds and is to be defined in terms of the categorial process. If reality is continuous but nevertheless yields concepts which are discrete, this is as it should be. There is no radical difference in kind between practical or scientific knowledge on the one hand and philosophic knowledge on the other, but only a difference of degree of generality.

VI

The present essay, in support of the view that the nature of the philosophic enterprise is the formulation, development and criticism of categories, has presented a sketch of a categorial scheme for the understanding of the categorial process. The field of philosophy as a whole is the understanding of experience in general; the special field of epistemology is the understanding of understanding. If understanding in general is accomplished by the development and application of categorial schemes, the understanding of understanding requires a special categorial scheme. To show that understanding can be understood by means of a categorial scheme is to present a crucial case in support of the contention that all understanding is accomplished through the development of categorial schemes.

The philosophic enterprise as a human activity takes its point of departure from the common sense world of the person embarking on it. Common sense, however, does not furnish the fundamental categories of epistemology. The development of epistemology is part of the philosophic enterprise, and its categories can be found only in pursuit of the enterprise. The philosopher in the order of time starts from common sense categories, but in the order of logic, he must start from categories at which he has arrived by means of logical analysis and criticism.

Chief among the categories offered in the present essay for an understanding of understanding is the concept of the flux of perceptual intuition. The flux is not a concept, but we can develop a concept of it. Thus, the term 'concept' names the second basic category of the system. A third basic category is named by 'relatedness'. Two more basic categories are 'property' and 'process'. The sub-categories of property are 'continuous' and 'discrete'. Concepts are derivatively related to the flux by the category of process, the sub-categories of which are 'analysis', 'abstraction' and 'generalization'. The relatedness of analysis to the flux is understood by means of another sub-category of process taken from analogy to the mathematical theory of the continuum — the Dedekind cut. The continuum can be analyzed or cut according to regularities, repetitions, places of high intensity and low intensity in its occurrence. These may be considered as sub-categories of 'continuous', and thus, as sub-sub-categories of 'property'. They are illustrated by the physical analogy to nodes in a continuum.

The relatedness between concepts is expressed by the sub-categories of relatedness 'inclusion' and 'exclusion'. The application of these sub-categories gives rise to the meaning of 'logic'. All categories are concepts, and all are related to each other in more and more complex logical ways as the cognitive process expands. The normative aspect of logic is expressed in the criterion of consistency of relations, but consistency itself is a matter of inclusion and exclusion and thus does not go outside the categorial scheme.

The knowledge of perceptual objects is made up of selections from the flux of intuited data — selections obtained by subsuming parts of that continuum under concepts obtained from cuts. A cut in the continuum sets a boundary, and concepts are always definitive. Thus, perceptual objects may be *understood* to be separate from each other, but they are not so given. The apprehension of discrete perceptual objects is a cognitive act, and the discreteness comes from the conceptual element in the perception of objects. A perceptual object is to be understood as the subsumption of a portion of the flux under a concept. Abstract concepts are obtained in the first place by abstraction and generalization from empirical concepts, but by repeated application of the processes of abstraction and generalization may be freed from the material or content of their origin. Thus, the concepts and conceputal schemes of pure logic and mathematics finally appear as criteria for our grasp of the flux; and understanding is attained.

As perceptual knowledge is to be understood as a categorization, philosophy itself is seen to be an uncommon extension of common sense. Philosophy goes beyond common sense in that it finds connection, relations, and thus, categories that are not obvious or are not already crystallized in ordinary language. In its critical capacity also, it goes beyond common sense, but it is hardly to be expected that it will develop categories that violate common sense or wholly reject it. The present view of the nature of philosophy does not, then, yield a common sense philosophy, for it is critical of common sense. Neither does it yield a philosophy that is violative of common sense, but rather one that is continuous with it. This seems to be as consistency demands.

Categories for the understanding of understanding have been proposed in this essay, but those proposed are not, probably, adequate to an understanding of 'truth'. Knowledge should not be confused with truth, although surely there is an important relation between them. The view that knowledge, in order to be genuine knowledge, must be true is itself due to a confusion of categories. All knowledge is fallible; that is, it may be erroneous. The categorial antitheses are 'true—false' or on the one hand and 'knowledge — ignorance' on the other. To identify falsity and ignorance requires a sleight-of-hand performance that is violative of the common sense from which all philosophizing starts. Truth and falsity are properties of propositions, and literally should be applied to nothing else. Knowledge, however, is the hypostatized resultant of a process. Knowing is a proper part of believing: all knowing is believing, but there are some beliefs that are not knowledge. Knowledge is that part of belief for which there is sufficient evidence; but 'sufficient' is a term of relative degree. Thus, knowledge itself is always present only in degrees. It is questionable what 'absolute knowledge' could mean.

It must be remembered that, according to the view presented in this essay, cuts can be made anywhere in the continuum. They can be made at different nodes or even at different "places" in the same node. Thus, alternate categorial schemes are always possible; and as any scheme is based on only a selection of what is there, no one categorial scheme can be said to be the correct one to the exclusion of all others. This caveat applies to the scheme sketched in the present essay. If I did not recognize that it does, I would be committing what Angus Sinclair calls "the epistemologist's fallacy."[11] This fallacy consists in assuming that my knowledge of the strictures and limitations to which I subject all other knowledge is somehow exempt from those strictures and limitations. Such an assumption has no logical justification, and to assert it is the sheerest of dogmatisms. Any philosophic system that presents what purports to be *the* list of categories is guilty of this fallacy, and hence, of this dogmatism.

Thus, I do not assert that the categories by means of which I understand understanding comprize the only rationally possible or even the "true" categorial scheme (whatever that would mean); but only that this categorial scheme is a reasonable one and that it supports the contention of the present essay that the nature of the philosophic enterprise is the formulation, development and criticism of categories by means of which the understanding of experience is made possible.

---

11 Angus Sinclair, *The Conditions of Knowing*, N. Y., 1951, p. 41.

THE NATURE OF ANALYTIC PHILOSOPHY

Paul Guerrant Morrison

SECTION I. INTRODUCTION

IN the widest sense, analytic philosophy is an activity,—the activity of discovering and exhibiting a structure of meaning fundamental to a progressively larger and more varied set of broad established concepts which reflect features of the world and of human discourse about the world. It is true, of course, that many efforts by analytic philosophers have been short-term affairs, aimed at discovering the meaning relations of smaller numbers of closely related concepts. In so far as such limited events contribute to the discovery and presentation of a larger meaning structure of the kind just mentioned, however, they belong to analytic philosophy.

When such an exploration of meaning is pursued far enough, one of its more important products is an organized body of statements revealing a system of meanings with room for the concepts of every special field of inquiry. In making it, however, the philosopher is not looking for factual or contingent truths. That is the business of scientists and men of practical affairs. The analytic philosopher is concerned, rather, with uncovering a general conceptual structure carried by expressions used in informative discourse, including those used to state contingent truths of every sort.

In the rest of the paper, the points just made will be presented in more detail and in a formal manner. Section II explains what is to be understood here by 'concept', after first defining 'proposition' and 'intension' in terms of designation, truth, falsity, and a physical part-whole relation. Structures of meanings, or "purviews", are then specified in Section III as certain class sums of propositions asserting nonfactual relations between concepts. Section IV goes on to develop the notion of one purview being "fundamental" to two or more otherwise "disparate" purviews. And this paves the way for a more exact characterization of the unifying nature of philosophic activity and of the role of analysis and construction in philosophy.

Section II. Propositions and Concepts

A. The Nature of Concepts

Since analytic philosophy is held to be concerned with the meaning structure of a set of concepts, it is essential that we specify what is meant here by the word 'concept'. Before setting out to define it, however, let us consider certain general features of concepts which must be accounted for in such a definition.

(1) *Concepts are objective or interpersonal.* That is, they can be grasped by more than one person, and one person can teach another to recognize, or to understand, them. On the other hand, the introspected events, or mental representations, which may be involved somehow in teaching or learning a concept are, as far as we know today, accessible only to the introspector. They are private and personal. It may be, of course, that we never communicate or grasp a concept without having such introspective or mental representations. But if we continue to regard concepts as objective and as communicable from person to person, it might be better, perhaps, not to identify them with something which is, for the present, at least, held to be inaccessible to more than one person.

(2) *Concepts are expressed by terms.* Almost everyone agrees that there is an intimate relation between concepts and terms. Certain linguistic expressions are commonly spoken of either as designating, or as expressing, or as standing for concepts. In accordance with the position taken above, however, we reject the traditional view that the concepts which terms designate or express are ideas in someone's mind. One method of avoiding this introspective view of concepts was proposed some time ago by Rudolf Carnap. In a more recent work, he explains it as follows:

> The term 'concept' will be used here as a common designation for properties, relations, and similar entities (including individual concepts . . . and functions, but not propositions). For this term it is especially important to stress the fact that it is not to be understood in a mental sense, as referring to a process of imagining, thinking, conceiving, or the like, but rather to something objective that is found in nature and that is expressed in language by a designator of nonsentential form. (This does not, of course, pre-

clude the possibility that a concept,—for example, a property objectively possessed by a given thing—may be subjectively perceived, compared, thought about, etc.)[1]

Carnap's view certainly satisfies our understanding of concepts as something objective. For the properties, relations, and functions studied by the sciences are what we usually appeal to when we are asked to characterize objectivity.

(3) *Concepts can be formed, modified, or discarded.* Carnap's interpretation of 'concept' as referring to a property, relation, function, etc., found in nature requires us to take account only of the linguistic expression and what it designates. Hence we might call it a *semantic* interpretation of 'concept'. And this interpretation is well suited for the works in which he uses it. For he is concerned primarily with semantical and syntactical problems in these works.[2]

Moreover, the semantic interpretation of 'concept' provides a simple explanation of the features of concepts just mentioned,—their objective nonmental nature, and their intimate connection with linguistic expressions.

The problem before us, however, is set in a somewhat broader context. For in considering analytic philosophy as a human activity concerned with concepts, we are widening the context of our discussion to include not only the linguistic expression and its designation, but also the user of the language. That is, we are adding a pragmatic dimension to the semantic and syntactic ones already implicit in Carnap's interpretation of 'concept'.[3]

An example or two will show what this involves. It is perfectly satisfactory, of course, in semantic discussions, either to say "The English expression 'circulatory system' designates the property Circulatory System", or to say "The English expression 'circulatory system' expresses the concept Circulatory System",—regarding one as practically synonymous with the other.[4]

---

1 Rudolf Carnap, *Meaning and Necessity*, p. 21. Although this interpretation of 'concept' is not the one chosen below, I am indebted to Professor Carnap for the liberating influence which this view has had on my own search for an interpretation of 'concept'.

2 See his *Introduction to Semantics*, (Cambridge, Mass., Harvard University Press, 1942), his *Meaning and Necessity*, (Chicago, Illinois, University of Chicago, 1947), and his *Logical Foundations of Probability*, (Chicago, Illinois, University of Chicago, 1950).

3 For the distinction between the three parts of the general theory of signs, or semiotic, into pragmatics, semantics, and syntactics, see C. W. Morris, "Foundations of the Theory of Signs", *International Encyclopedia of Unified Science*, I, No. 2, (Chicago, Illinois, University of Chicago, 1938).

4 In *Logical Foundations of Probability*, pp. 7-8, Carnap says: "If I want to speak about a concept (property, relation, or function) designated by a word, I sometimes use the device of *capitalizing* the word . . . For example, I might write 'the relation Warmer' . . . similarly, I shall sometimes write: 'the property (or concept) Fish' (instead of 'the property of being a fish'; 'the property (or concept) Red' (instead of 'the property of being red' or 'the property of redness'), and the like."

On the other hand, we would not be as willing to regard "Harvey formed the concept of being a circulatory system" as practically synonymous with "Harvey formed the property of being a circulatory system." For in the sense that Harvey formed the concept, whatever it was that he formed arrived on the scene much later than blood circulation did. That is, Harvey did, in some sense, form a *new* objective concept. And in that sense, the exemplification in nature of the concept Circulatory System began much later than that of the property Circulatory System. It is possible, of course, to retain the semantic interpretation of 'concept' here, by translating "Harvey formed the concept Circulatory System" into "Harvey was the first known man to designate the objective property (or objective concept) Circulatory System in any known language," —thus regarding the expression 'concept formation' as an elliptical one requiring careful paraphrase in technical usage.

On the other hand, it is possible to give a different interpretation of 'concept', which retains all the advantages of the semantic interpretation, but which also makes possible a simpler, more direct account of such operations on concepts as forming, using, discarding, and borrowing them, and of such properties of concepts as being old or new, fruitful or inadvisable, and customary or uncustomary. Such a *pragmatic* interpretation of 'concept' will be developed formally below. Roughly speaking, it will construe a concept as a certain kind of omnilingual class of sign-events[5] with the same meaning structure.

In this sense, the concept Hot Water, for example, comprises the class of all utterances or inscriptions of the English words 'hot water', the French words 'eau chaude', the German words 'heisses Wasser', etc. In other words, the concept in question is the class of all sign-events,—past, present, and future—, of any expression in any language which designates hot water in terms whose elements can be used elsewhere separately to designate hot things which are not water and aqueous things which are not hot. Each sign-event, on this view, is both an instance of an expression in some one language and an instance of a concept whose use is not necessarily confined to any one language. Thus, while the semantic interpretation construes a concept extensionally as an entity found in nature, the pragmatic interpretation takes the concept intensionally as an omnilingual class of sign-events (with like meaning structures) which designate such an entity.

---

5 "The word 'sign' is ambiguous. It means sometimes a single object or event, sometimes a kind to which many objects belong. Whenever necessary, we shall use 'sign-event' in the first case, 'sign-design' in the second." Carnap, *Introduction to Semantics*, p. 5.

The proposed pragmatic interpretation of 'concept' has two main advantages.

(1) It enables us to distinguish between a philosophic and a linguistic concern with language. For while the linguist and the philosopher may sometimes deal with the same utterances and inscriptions, the linguist is concerned with them as instances of terms or expressions occurring within a given natural language, while the philosopher is concerned with them as instances of concepts. And concepts, in the pragmatic interpretation given below, are omnilingual entities.

In another sense, the philosopher, too, is concerned with the terms of some one language. For, like everyone else, he must express himself in one or another of the natural languages. And in doing so, he must occasionally refer to terms occurring in such a language. But even when he is doing this, the philosopher is interested in these terms more for the concepts they express than for the language to which they belong.

Again, while it is true that the analytic philosopher in one sense produces his own language vehicle for exhibiting the conceptual system which he discovers, this product is not a language in the sense in which the linguist is concerned with languages. It is rather a language system,[6]—a personally manufactured system of rules which the philosopher develops and modifies deliberately as a tool for a single purpose,—the exhibition of the meaning structure of a system of fairly exact concepts. It is probably correct to say that, while the analytic philosopher does concern himself professionally with the expressions occurring in such language systems, the contemporary linguist has no professional interest in studying them.

(2) The pragmatic interpretation of 'concept' helps us give a simpler (although not a more objective or complete) account of man's intellectual activity than the semantic interpretation does. It is true that Carnap intended what I here call the semantic interpretation of 'concept' only as an informal characterization, and not as a definition. Thus he was not aiming at an interpretation calculated to make 'concept' a univocal term wherever it is used in analytic philosophy.[7]

On the other hand, if an interpretation like the semantic one were formalized, it could not, like the pragmatic one, be used in clarify-

---

6 "Die Art der semantischen Analyse ist anders, wenn es nicht um die empirische Untersuchung einer historisch gegbenen Sprache handelt, sondern um den Aufbau einer künstlichen Sprache. Anstatt ‚Sprache' sagen wir in diesem Fall oft ‚Sprachsystem', um zu betonen, dass es sich hier nicht um eine natürliche Sprache, sondern um ein System von Regeln handelt." R. Carnap, *Einführung in die symbolische Logik*, (Vienna, Austria, Springer-Verlag, 1954), p. 71.

7 Cf. Carnap, *Meaning and Necessity*, p. 22.

ing such expressions as 'concept formation', 'new concept', 'broadened concept', 'concept modification', and 'fruitful concept', without our using an interpretation of 'concept' in the latter expressions which differs from the semantic one.

Thus the sense of 'concept' in which Radioctivity is a new concept is not the same as that in which the concept Radioactivity is the property of being radioactive. For while the new concept, in this case, has been exemplified or expressed (by human beings on this earth) for less than one hundred years, the property of being radioactive has been exemplified for millions of years.

Again, the sense in which the older concept Fish (which included reference to whales, seals, etc.,) has been modified or replaced by a newer concept Fish* (reflecting current usage among biologists),[8] is not the sense in which the older and newer concepts of fish are properties of physical things. For the exemplification of the one property has not given place to that of the other in the waters of the world. It is rather that utterances and inscriptions of the expressions 'fish', 'Fisch', 'poisson', etc., are no longer being used to denote just any animal that lives in the water, but only aquatic animals which also have certain further characteristics.

It is thus the designative features of certain expressions, rather than the exemplification of what they designate, that is modified or replaced. In other words, even after a concept is discarded, the trait which it represents may continue to be found in the world. We simply feel this trait, or instances of it, too unimportant to talk about any longer. As indicated above, much of this can be (and has been) explained indirectly in terms of the semantic interpretation. But the pragmatic view allows us to use the same interpretation of 'concept' in the statement 'Concepts are objective, nonmental, and expressed by designators of nonsentential rank' as that used in the expressions 'concept formation', 'fruitful concept', 'new concept', 'concept modification', and 'customary concept'.

B. THE DEFINITION OF 'CONCEPT'

In the view presented below, concepts and propositions will be construed as kinds of meanings. And a "meaning" will be understood as a certain kind of omnilingual class of sign-events each of which designates the same entity. So conceived, meanings may be classified as "notions", or extensional meanings, and "intensions", or intensional meanings. Of these two, the extensional meaning, or

8 Cf. Carnap, *Logical Foundations of Probability*, p. 5.

notion, is the more general, in the sense that every intension turns out to be a subclass of some extensional meaning. Hence the definition of 'intension', below, is built up from the definition of 'notion'. Propositions can then be construed as intensions of certain kinds. And finally, a concept can be taken as a certain kind of intensional segment of a proposition.

A clear distinction between intensional and extensional meaning is basic to the interpretation of the words 'proposition' and 'concept' given below. Such a distinction was made by C. I. Lewis, when he asked whether pairs of expressions like '2+2' and '4' have the same meaning. In order to give a satisfactory answer, he made a distinction like the one needed here, to the effect that '2+2' has the same "holophrastic" or overall meaning as '4', but a different "analytic" meaning.[9] A notion, or extensional meaning, in the sense of the present paper, corresponds roughly to Lewis' holophrastic meaning. And an intensional meaning, or intension, in our sense, corresponds roughly to his analytic meaning. For in our sense, while the symbols '2+2' and '4' express the same notion, they express different intensions. Moreover, '2+2' and '4' are, for us, subclasses of the notions and intensions which they express. That is, we shall construe expressions as subclasses of their own meanings.

In our sense, then, while '4', '3+1', and '2+2' all express the same notion, they have, or express, different intensions. In the same sense, '4', 'four', 'vier', 'quatre', etc., all express the same notion and express, or have, the same intension. And so do '3+1', 'three plus one', 'eins und drei', 'un et trois', etc. Our immediate problem is to find a way to define what we mean by 'notion' and 'intension' in such cases without employing introspective terms in the definitions. A soluton of this kind will be attempted shortly below.

In working out the answer to this problem, and throughout the rest of the paper, we shall need to express our results in a formal language system. In order to distinguish the official statements of this system from the word commentary which accompanies them, let up adopt the device of italicizing and following each one of them with a literal and numerical label, showing whether it is an axiom (A), a theorem (T), a definition (D), or an announcement of a primitive (i.e., undefined) descriptive (i.e., nonlogical) expression (P). Thus '(A1)' will indicate the first axiom; '(T5)', the fifth theorem; '(D10)', the tenth definition; and '(P2)', the second undefined descriptive expression of the system. And finally, when-

---

9 Cf. C. I. Lewis, *Analysis of Knowledge and Valuation*, (LaSalle, Illinois, Open Court, 1946), pp. 85ff.

ever a new descriptive expression is being used for the first time
in one of the official statements, it will be capitalized.

To start with, then, let us deal with the formal specification
of our concept of notion, or extensional meaning. We can define
'notion' with the help of only one nonlogical primitive, 'x DESIG-
NATES o'[9a] (P1), where x is any extended space-time thing or event,
and where o is any entity whatever. If null entities of appropriate
kinds are admitted in our conceptual scheme, then we may stipulate
that *any x which designates some entity, designates that entity
alone* (A1). With this reservation, *a NOTION, f, may be defined
as a maximum class of extended space-time individuals each of
which designates the same entity* (D1).[10] In this sense, for example,
the first utterance of 'two plus two' in a grade school in Ohio on
April 2, 1958 and the last inscription of '3+1' on a blackboard in
Wiesbaden, Germany on that same date in 1951 are instances of the
notion Four.

'3+1' and '2+2' both express the same notion. But they do so
in different ways. The traditional manner of expressing this difference, with which we agree here, is to say that they have the
same extension, but different intensions. How may we explain the
different ways in which the sign-events of '3+1' and those of '2+2'
designate the same entity? One answer might be to say that the
meaning of each expression has "phases" which do not belong to
the meaning of the other. Thus the '3' in '3+1' expresses a phase of
meaning not expressed by any element of '2+2'. And either of the
expressions, '2' in '2+2' expresses a phase of meaning not exhibited
by any element of '3+1'. In contrast, the meaning of '+' is a phase
both of the meaning of '3+1' and of the meaning of '2+2'.

If meanings are construed as classes of sign-events, how may
we specify such phases of meaning in those terms? An answer to
this question can be given shortly, if we first introduce another
nonlogical primitive, 'x is PART of y' (P2). 'x is part of y' is to be
understood here in the sense that both x and y are extended space-
time things or events, so that x is an extended part of y. Moreover,
it will greatly simplify the remaining official statements of the
formal system if we take identity of extended individuals as a
limiting case of the part-whole relation. That is, instead of leaving
off using the expression 'part of' when x is almost, but not quite,

---

9a Throughout the paper, we shall use the letter 'o' for entities of all kinds;
the letters 'x' and 'y', for individual things or events; the letters 'a' through 'i', for
classes of individuals; and the letter 'k', for classes of classes.

10 More formally, '*f is a NOTION*' means that there is an entity, o, such that any
x designates o if, and only if, x is an instance of f (D1A).

all of y, we broaden the meaning of 'part of' to the limit, by an axiom saying that *every extended individual is part of itself* (A2). Let us now turn again to consider phases of meaning. *Sometimes two notions, h and i, are so related that some of the sign-events of h are parts of sign-events of i. In that case, the class, f, of all such sign-events of h is a PHASE of the class, g, of all sign-events of i of which they are parts* (D2). Thus, for example, if f be a subclass of the notion Three, and if g be a subclass of the notion which includes Three Plus One, where f is a phase of g, then the sign-events which exemplify this meaning-phase relation are not restricted to utterances or inscriptions of the English expression 'three plus one', but also include sign-events of the French expression 'trois et un', the German expression 'drei und eins', etc. For although the meaning-phase relation holds between classes of sign-events, those classes are, as a rule, larger than the expressions of any one language which can express the meanings in question. For they are meanings shared by many languages. Since every sign-event is part of itself, it follows from the definition of 'phase' that *every phase of anything is a phase of itself* (T1). It also follows that *every notion is a phase of itself.* (T2).

Moreover, since a phase is always a maximum class of notional parts, and since notions themselves are phases, it turns out that *any class, f, is a MEANING, whenever f is a phase of some g* (D3).

The terms 'meaning' and 'extensional meaning', or 'notion', have been *defined* above in terms of 'designates' and 'part'. But so far, we have only *characterized* the meaning of 'intension'. The point has now been reached at which we can explain how certain equivalent expressions of the same notion, such as '3+1' and '2+2', have different intensions. For we can now state what it means for the meanings which they express to have different phases of meaning. But if we are to *define* 'intension', we must do more than contrast meanings which differ in intension. And this step can now be accomplished. For if the sign-events which comprise a notion are subdivided into different intensions because of meaning-phase differences, an intension will be a maximum class of sign-events which do not manifest such differences. In terms developed above, *any class of sign-events, f, is an INTENSION, whenever f is a meaning, and every phase of every subclass of f is also a phase of f itself* (D4).

Thus, for example, 'three plus one', 'drei und eins', 'un et trois', '3+1', and '1+3' express, or have, the same intension. For every meaning-phase expressed by any subclass of the sign-events belonging to one or more of those expressions is a phase of the overall meaning expressed by each full expression. In contrast, '3+1' and '2+2' do not have the same intension, because the meaning-phase expressed by the '3' in '3+1' is not expressed by any element of '2+2', and hence, is not a phase of the larger class of sign-events of which both '3+1' and '2+2' are subclasses.

In terms of developments to this point, we may now lay down a theorem stating that *every intension has at least one phase* (T3).

We have now reached a point at which propositions and concepts may be specified in systematic terms. For since propositions, like other meanings, are construed here as omnilingual classes of sign-events, we may understand *any class, f, to be a PROPOSITION, whenever f is an intension which is true or false* (D5). If some readers object to calling anything other than an expression true or false, it should be noted that an intension, in this system, is always a class sum of univocal expressions from one or more languages. And if philosophy and logic are not particularly concerned with cultural or natural-language differences in the phonetic or graphic structure of sign-events with the same meaning structure, this extension of meaning is warranted.

Concepts may be defined in terms of propositions. Yet while every concept occurs in some proposition, or is expressed by an element of some sentence, concepts are not usually construed as whole propositions or as meanings of whole sentences. On the other hand, to deny that a proposition might be an element of a concept, would be to err in the other direction. For then we should have to deny conceptual status to such meanings as Utterance of 'I pronounce you man and wife'. If we may call *a class, f, a DETAIL of another class, g, when every instance of g has an instance of f as part* (D6), we may still define 'concept' in terms of propositions, but in a way which avoids both the extremes just mentioned. Thus, *an intension, f, is a CONCEPT, whenever f is a nonpropositional detail of some proposition* (D7).

The concepts whose meaning relations the analytic philosopher explores and exhibits are not just any concepts, however, but only those which play some significant role in man's quest for knowledge. For obviously, if we may form a concept simply by using language to designate some entity which has not yet been spoken of, we can spin out trivial objective concepts by the yard. For example, we can designate the property of being a bulldog in a triangular park in which May Day parades originate when there are cumulus clouds overhead, the relation of being sixteen years older than one or more persons who were born on some rainy Tuesday or other, etc. Clearly, only concepts with certain additional features are worthy of sustained consideration. We shall not attempt a systematic survey of these features here. But whatever they are, it is obvious that many of the established concepts of modern science must have them. This can be seen from the fact that these concepts have participated in a progressively more effective manner in man's struggle to obtain knowledge of, and control over, his environment. It is for this reason that the analytic philosopher most often chooses established *scientific* concepts as the raw material for his conceptual exploration.

SECTION III.   PERSPECTIVES AND PURVIEWS

If analytic philosophy involves discovering a conceptual structure fundamental to all fields of inquiry, it is important to indicate the nature of such a structure. In the view given below, a conceptual structure, or "purview", as we shall call it, will be understood as a class sum of related "meaning assertions", or propositions asserting certain nonfactual connections between concepts. But since the usual sort of declarative passage expresses some propositions with factual content, we shall first develop the more general notion of a "perspective",—or class sum of related propositions of any kind—, and then specify the purview as a perspective consisting entirely of related meaning assertions.

Propositions can be related in various ways, however; and we need to show the particular manner in which they are related in a perspective. The basic relation involved is that of "convergence" between propositions,—a relation which binds them when they have a meaning phase in common. More precisely, *a proposition, f, CONVERGES with another proposition, g, when some h is a phase both of f and of g* (D8). For example, the proposition expressed by the statement 'Two plus one equals three' and that expressed by the statement 'Trois et quatre font sept' converge, because the concept Three is a meaning phase of each proposition.

For a declarative passage to give a unified point of view, it is obviously unnecessary for every pair of propositions expressed by it to converge. The most that can be demanded is that each such pair be linked by a finite number of convergence steps,—that is, for any such pair of propositions, f and g, that f either converge with g, or converge with some h which converges with g, or with some i converging with some h which converges with g, etc. In these terms, *any class of sign-events, f, will be called a PERSPEC-*

TIVE, when *f* is a finite logical sum of propositions each of which is linked by one or more convergence steps to every other (D9).

As indicated above, a conceptual structure, or "purview", is to be construed here as a perspective made up exclusively of "meaning assertions"—i.e., of propositions which declare certain nonfactual relations between concepts. Now since a meaning assertion is to be taken as a proposition, and hence as an intension, it is obvious that the concepts which it relates will be phases of the proposition itself. Moreover, if we agree to call *any class, f, DISTINCT from another class, g, when no instance of f has a part in common with any instance of g* (D10), it is clear that the phases between which the meaning relation is asserted will be distinct.

For example, in the proposition All crows are black, understood as a meaning assertion,—i.e., understood as a proposition declaring a nonfactual relation between a concept expressed by 'crow', 'corbeau', 'Krähe', etc., and one expressed by 'black', 'noir', 'schwarz', etc.—, the concept of crows in question is distinct from the concept of black. For no sign-event of the proposition has a part expressing its concept of black overlapping a part expressing its concept of crows. Hence its concept of black can not be a meaning phase of its concept of crows. And yet, there is a broader sense in which we might like to say that the concept of black in question is a meaning feature of the concept of crows in this proposition. In fact, we want to call the proposition a meaning assertion precisely because it represents the color concept as a constant meaning feature, or "aspect", of the concept of crows expressed there.

If the relation we seek is broader than the meaning-phase relation, however, it is still akin to it. Thus, for example, since every sign-event of the concept Father-of-a-Parent-of has a sign-event of Parent-of as a physical part, the concept Parent-of is a phase of the more complex concept. But while the concept Parent-of is a phase of the concept Father-of-a-Parent-of, it is not a phase of the equivalent concept Grandfather-of. On the other hand, in the sense that every grandfather is a father of someone's parent, the concept Parent-of is a meaning feature, or "aspect", of the concept Grandfather-of. Perhaps, then, if we first specify the meaning of 'equivalent' in this context, we can define 'aspect' in terms of equivalence and the meaning-phase relation.

'Equivalence' can, in fact, be defined quite simply in terms of designation. For *any class (of sign-events), f, is EQUIVALENT to another such class, g, when there is an entity, o, such that every sign-event of f and every sign-event of g designates o* (D11). The concept Two, for example, is equivalent, in this sense, to the concept Even Prime Number. Again, the concept Two and the concept Even Prime Number are both equivalent to the notion Two, which includes them. In fact, one of the useful features of this notion of equivalence between classes of sign-events is that sometimes such a class, f, is equivalent to one of its subclasses, g, or is equivalent to a class, h, of sign-events of which f itself is a subclass. Moreover, it follows from the definition that *every class of sign-events is equivalent to itself* (T4).

Returning to the example given earlier, we may recall that, while the concept Parent-of is not a phase of the concept Grandfather-of, it is a phase of the concept Father-of-a-Parent-of, which, in turn, is equivalent to the concept Grandfather-of. From this example, it might appear that we could define 'f is an aspect of g' by saying that f is a phase of an equivalent of g. That would be too restrictive, however.

It is true, of course, that we would like to be able to call one concept an aspect of another concept. But since there are other kinds of meanings besides concepts, it would seem more appropriate to define 'aspect' as a relation holding between meanings of any kind. It would be nice, for example, to be able to call the notion Two, to which the concept Two and the concept One Plus One belong, an aspect of the notion Four, to which the concept Four, the concept Three Plus One, the concept Two Plus Two, etc., belong. And since a class of sign-events can be equivalent to one or more of its subclasses, we can specify the view that the notion Two is an aspect of the notion Four by saying "the notion Two is equivalent to a phase of an equivalent of the notion Four". More generally, *any f may be called an ASPECT of some g, when f is equivalent to a phase of an equivalent of g* (D12).

It follows from the definition of 'aspect', that *every phase of any f is also an aspect of f* (T5). And from the reflexive nature of the meaning-phase and equivalence relations, it follows that *every aspect of anything is an aspect of itself* (T6).

In defining 'aspect', we have uncovered an internal relation which is essential to the meaning assertion. For as we have seen, such an assertion is a proposition which declares a certain nonfactual relation to hold between two of its mutually distinct phases. And this relation consists in one of these phases being an aspect of the other. Thus, if we take the proposition All Crows are Black as truly asserting that its concept of crows is an aspect of its concept of black, then that proposition is a meaning assertion. More generally, *any proposition, f, is a MEANING ASSERTION, when f has some distinct phases, g and h, such that f is true if, and only if, g is an aspect of h* (D13).[11]

At this point, the reader may object that the proposition expressed by the statement 'All crows are black' does not necessarily assert a meaning relation, but may assert a factual relationship. And since closer consideration reveals that this is a difficulty shared by a vast number of declarative statements, the objection might continue, there may be no way to tell whether or not a statement is factual. And if that is the case, the distinctions made above will be inapplicable and quite useless. In reply to this, it must be admitted that it is sometimes hard to tell whether a certain statement in everyday discourse is meant to assert a fact or to declare a meaning relation. Still, this difficulty is not insuperable.

For let us suppose that we find the sentence 'All crows are black' in a conventional prose passage dealing with birds. Did the writer mean that every offspring of black crows will, in all probability, turn out to be black? If so, he was making a factual claim in asserting the statement. On the other hand, it might be that he would not consider the further statement 'Whitey is the albino offspring of a pair of black crows' as evidence of the falsehood of 'All crows are black'. That is, he might answer that Whitey is perhaps crow-like, but clearly not a crow, since only black crow-like birds are crows. Thus we would eventually learn that he had probably asserted 'All crows are black' as a meaning statement. In the absence of such a clarifying discussion, however, we might not know whether the writer intended 'All crows are black' as a factual statement or as a meaning statement. For nothing *per se* about the form of a nontautologous descriptive statement shows which way it is intended.

Moreover, this example illustrates a kind of vagueness, if we assume that 'black crow' designates the same class, g, of individuals on either interpretation. For if 'All crows are black' is a meaning statement, then 'crow' in that passage designates g and nothing else. But if it is a factual statement, then g is only a large proper subclass of what 'crow' designates there, since Whitey would not be a member of g, and yet would be considered a crow whose existence would then show 'All crows are black' to be false.

---

11 From this definition, it will be clear that all meaning assertions are analytic. Some analytic propositions, however, are not meaning assertions. For example, while the propositions expressed by '1 = 1' and by 'If p, then p' are not meaning assertions, they are analytic.

In the terms of our system, since the entities designated differ in this way, the concepts expressed in 'all crows are black' taken as a meaning statement will not necessarily be the same as those expressed by 'All crows are black' taken as a factual statement. Because of the large overlap in the classes designated by sign-events of the concepts of crows in the two cases, we might call those concepts "variants" of one another.

The difficulty in determining whether a declarative statement expresses a factual proposition or a meaning assertion becomes negligible, however, when we turn from passages in a natural language to passages in formal language systems. For it is customary in such systems to label the member statements to indicate their capacity.[12]

Now that the character of meaning assertions is clear, we may, at last, specify the nature of "purviews",—those complex meaning structures, devoid of all factual content, with which the analytic philosopher is most directly concerned. For the philosopher has always been interested in conceptual systems primarily, or, in our terms, in perspectives. But as more and more of the factual disciplines have been withdrawn from the province of philosophy during the last two thousand years, his more particular concern with those pure systems of meanings, here called purviews, has become increasingly evident. In the remainder of this paper, we shall consider the analytic philosopher's concern with purviews in greater detail. Before doing that, however, we must first define 'purview'. And in terms developed above, *any perspective, f, is a PURVIEW, when every proposition included in f is a meaning assertion* (D14).

One simple relation between purviews is of special importance to the analytic philosopher,—the relation of "consonance". (*One purview, f, is CONSONANT with another, g, when every aspect of g is an aspect of f* (D15).) The relevance of this notion will become obvious in the next section, when we specify what it means for one purview to be a "basis" for another.

SECTION IV. PHILOSOPHIC ACTIVITY

As intimated above, analytic philosophy is conceived here as the activity of discovering and exhibiting a unifying conceptual structure,—a structure fundamental to the broad established concepts of every special field of cognitive inquiry. Using terms just

12 For perhaps the first suggested way of doing this, see Rudolf Carnap, "Meaning Postulates", in *Philosophical Studies*, III, (1952), pp. 65-73.

developed, the analytic philosopher seeks a broad purview which is fundamental to a number of less comprehensive purviews that have no unifying basis *per se*. In what follows, we shall first determine the nature of such a fundamental structure, and then specify the character of the activity through which it is discovered and presented.

A. FOUNDATION STRUCTURES

Since unification involves difference, our first task is to indicate what it means for two purviews to be conceptually "disparate." Then we can show what we understand by a purview unifying,— or providing a common basis for—, disparate purviews. And afterwards, the notion of a broad fundamental purview,—a conceptual structure providing a common basis for many limited purviews—, may be specified.

Two purviews will be construed as disparate, on the present view, when neither of them affords a conceptual "basis" for the other. And in this context, one purview, f, will be counted a basis for another purview, g, when f contains a partial or total "reduction" of the unreduced, or "primitive", meaning phases of g. Then, whenever a pair of purviews are disparate, a third purview will be understood as "fundamental" to them when it is a basis for each.

In formalizing the view just outlined, our first step will be to specify the meaning of 'reduction', in the sense in which a *complex* meaning constitutes a reduction of a *simple* one. And since meanings were defined above as phases, a simple meaning, or "simple phase", can be defined easily in our system. *Any f is a SIMPLE PHASE of some g, when no phase of g other than f is a phase of f* (D16). In this sense, for example, the concept expressed by utterances of 'city' is a simple phase of the propositions of a purview in which it occurs. On the other hand, while the concept expressed by sign-events of 'large city' may be a phase of such propositions, it is not a simple phase of them. For it has the concept of city in question as a partial phase. The concept Large City, then, is a "complex phase" of the propositions in which it occurs. More generally, *any f is a COMPLEX PHASE of some g, when f is a phase of g other than a simple phase of g* (D17).

We can now state formally what it means for one phase of a larger meaning context (e.g., of a purview), h, to be a "reduction" of another phase of h, in the sense, for example, in which the concept Even Prime Number is a reduction of the concept Two in some arithmetical purview. For the former concept is a complex phase, to which the later, a simple phase in the same purview, is equivalent. More generally, *any f is a REDUCTION of some g in some h, when f is a complex phase, and g, a simple phase, of h, where f is equivalent to g* (D18).

In the view we are developing here, the notion of a "fundamental" or unifying purview is to be specified in terms of one purview, f, being a "basis" for another, g, in the sense that f reduces some or all of the unreduced, or "primitive", meanings of g. Consequently, before we can define 'basis' in that sense, we must first clarify the notion of a primitive, or unreduced, meaning phase. Clearly, a meaning that is unreduced in one purview might be reduced in some other. Hence a notion of absolute irreducibility would be of little use in our system. The notion of a meaning being unreduced, or "primitive", *in a given purview*, however, will be quite helpful in the definitions of 'basis', 'disparate' and 'fundamental' below. Let us say, then, that *any f is a PRIMITIVE PHASE of some g, when f is a simple phase of g for which there exists no reduction in g* (D19).

If one purview, f, is to be counted a conceptual basis for another, g, then every aspect of g must be an aspect of f. That is, f must be consonant with g. But more than this is required. For every purview is already consonant with itself. And the purviews we are now looking for would presumably contain elements of some of the unreduced meanings of the purviews whose bases they form. Thus, if f and g are both arithmetical purviews where f is a conceptual basis for g, then if a numerical concept of two is unreduced (i.e., primitive) in g, f might contain a concept of even prime number,—a reduction of that concept of two. In formal terms, then, *a purview, f, is a BASIS for another, g, when f is consonant with g; when some primitive phase of g has a reduction in f; and when every primitive phase of g either is a primitive phase of f, or has a reduction in f* (D20).

The partial purviews which arise in various kinds of human inquiry quite often lack a common conceptual basis *in se*. In most cases, neither of a given pair of such limited purviews affords a conceptual basis for the other. When this is the case, we shall call each such purview "disparate" from the other. In systematic terms, *a purview, f, is DISPARATE from another purview, g, when neither of them is a basis for the other* (D21).

Since analytic philosophy is construed here as an activity of discovering and exhibiting a broad conceptual structure, or purview, which unifies a number of less comprehensive purviews, we must next determine what is meant by one purview being "fundamental" to two or more others. In looking for fundamental purviews, however, the analytic philosopher need only find a common basis for two disparate purviews at a time. For whenever he succeeds, this common basis plus the two disparate purviews in question comprise a new purview, which, as a next step, can be linked in the same way to yet another purview with which it, in turn, is disparate. Consequently, in this system, *a purview, f, will be called FUNDA-MENTAL to two others, g and h, when f is a basis for g and for h, where g is disparate from h* (D22). In confirmation of the point just made above, we lay down an axiom stating that *whenever f is fundamental to g and h, the logical sum of f, g, and h is a purview* (A3).

It also follows from the foregoing, that *if f is fundamental to g and h, then f is a basis for the logical sum of g and h* (T7). Hence, for example, by first finding a basis, c, fundamental to two purviews, a and b; then finding a basis, e, common to c and a further purview, d; and next, a basis, g, common to e and another purview, f; etc., the philosopher can move steadily to more and more comprehensive bases for the established concepts whose unification he is seeking. Let us call any comprehensive basic purview of this kind, unifying the members of a class, k, of limited purviews, a *"foundation structure"* for k.

It may have occurred to the reader that any one purview can have more than one basis, and hence, that there may be more than one purview, f, fundamental to a given pair of disparate purviews, g and h. Moreover, by the same token, there will be a plurality of possible foundation structures for the same comprehensive class, k, of limited purviews whose unification the philosopher is seeking.

These possibilities give rise to a methodological question for the analyst. For on what principle will he choose between possible alternative bases for any purview in working towards a foundation structure? One answer can be formulated in terms of the comparative "richness" of alternative bases for the same purview. For of any two bases for the same purview, f, which reduce different numbers of primitive phases of f, we may call that one the *"richer"* which reduces the greater number of those primitive phases. Thus at least a partial answer to the question is available when differences in richness exist. For in such cases, we may always choose the richer basis at each step in working towards a foundation structure.

On the other hand, it may be that at one or more steps in the development of a foundation, all available alternative bases reduce the same number of primitive phases of a purview pair to which they are fundamental. In that case, more than one foundation structure for the same initial set of purviews may result from the analysis. And this contingency gives rise to a second question of procedure. By what criterion may we select the most satisfactory of a number of equally rich alternative foundation structures for the same set of limited purviews? Once again, only a partial answer is offered here, covering those cases in which the alternative foundations have different numbers of primitive phases. For then, the most satisfactory foundation structure for the set of purviews in question is that which, itself, has the least number of primitive phases.

B. ANALYSIS AND CONSTRUCTION

The philosopher may be said to *analyze* a comprehensive and varied set of established meanings, when he is first in expressing a new foundation structure for the class of limited purviews made up of those meanings. And such an initial expression of a foundation structure is a creative act. For it involves the first expression, i.e., the *"conceptualization"*, of many hitherto unexpressed relations between aspects of the established concepts under analysis.

From another viewpoint, this same act, since it involves such original expression, is also an act of *construction*. For an expression is regarded here as a maximum class of physically resembling sign-events in some one language each of which designates the same entity. And when a previously unexpressed relation between aspects of established concepts is first expressed, a new expression, designating for the first time the corresponding relation holding in the world, is constructed. Philosophic activity, then, is a simultaneous process of analysis and construction. And viewed either way, it is a creative process.

While creative, however, philosophic activity is not arbitrary. For in engaging in it, the philosopher is not free to construct just any expresson he pleases. Nor is he at liberty to conceptualize any relation he chooses. For although a great many relations, unexpressed before an analysis begins, may hold within a set of concepts to be analyzed, the philosopher, in performing his analysis, is limited to discovering, i.e., to being the first to express, concepts from just that totality of relations. And thus, he is also limited in his construction to coining terms which express just those hitherto unexpressed relations.

Moreover, the philosophic enterprise is externally determined in another way. For both what the philosopher can discover and what he can exhibit depend on important features of the world and of human discourse about it. And since the latter features tend to reflect the former, there is little that the philosopher can do directly, as a philosopher, to alter the varied totality of broad established concepts which furnish the raw materials for his comprehensive analysis. Furthermore, because of new scientific advances and marked changes in man's environment, this totality of basic concepts is continually changing. And when the philosopher reflects a basic change of this kind in a new language system, there results a new foundation structure containing a great many variants of the concepts of the earlier foundation.

Because of those perpetual changes, the process of philosophizing is never completed. In fact, we can not rationally hope for its completion. For analytic philosophy, like the conceptual structures it produces and the world whose features these structures partially reflect, is essentially dynamic.

WILMON H. SHELDON'S PHILOSOPHY OF PHILOSOPHY

Andrew J. Reck

I

"PHILOSOPHY is doubtless the noblest and, at the same time, if we judge by overt results, the most futile of human enterprises."[1] Although on the contemporary scene allegations of the futility of philosophy outnumber and overwhelm avowals of its nobility, to him who is actively engaged in philosophy its nobility should need no recommendation. Whether his task be the analysis and criticism of received social, scientific and religious statements, or the reconstruction of principles and categories indispensable to correct theories or sound policies of action, or, highest of all, the creative preoccupation with the expression of a new vision of reality in a comprehensive philosophical system, the philosopher need not doubt the intrinsic value of his work. All the commonplaces concede that putatively insight, wisdom, truth are the veritable rewards of philosophy. On the other hand, some thinkers, born in this age of anxiety, have amended the commonplaces to imply that even if the insight were a vacuous gaze into sheer nothing or, worse still, the penetration into unutterable evil and ugliness, even if the truth prove to be an ineffectual wail in the face of cosmic meaninglessness, then the philosopher would at least have the value of his gaze and the significance of his truth. W. M. Urban pointed out several decades ago that, should the worst be true, the philosopher in discovering it so would have added the values of truth and meaning to the world.

At its best perhaps, as W. P. Montague has plausibly argued, philosophy is vision. It is a synoptic comprehensive survey of all

---

1 Wilmon H. Sheldon, *God and Polarity: A Synthesis of Philosophies* (New Haven, 1954), p. 3.

111

that is, and so Olympian an intention cannot fail to be noble. The most theoretical of cultural endeavors, philosophy is nonetheless the most practical. Spurred by man's essential inquisitiveness, philosophy, seeking a total system of all knowledge, proffers when successful the most complete view of the cosmos and man's place in it. Embracing all the areas of human activity — politics, history, art, science, religion, and the common life of plain men, philosophy is satisfied with nothing less than the most adequate and coherent integration of all these materials. As a science of first principles, it subordinates and rules what it integrates. Encompassing all realities within its systematic structure and ranking them in a hierarchy of values, it governs the practical. As J. K. Feibleman has shown, cultures are the embodiments of covert philosophies, and individual human conduct is the public expression of sub-conscious philosophical beliefs. The philosopher takes as one of his multifarious roles that of the specialist who explicates the cryptic, ferrets out the presuppositions, uncovers the "implicit dominant ontology," formulates and criticizes the postulates of public and private action.

The nobility of philosophy, then, springs from its practical as well as its theoretical intention. In appreciating and stating philosophy's dual role in human life, no philosopher in recent times has surpassed Wilmon H. Sheldon, Clark professor of philosophy emeritus at Yale. Acknowledging the rights of the pure contemplation of theory, Sheldon has consistently emphasized that theory issues practically in the governance of conduct. Forty years ago he wrote:

> There is then one, and only one, of all our human wants, whose satisfaction goes far toward satisfying the rest; that is, the need of a knowledge of the character, on broadest lines, of our universe. Such a knowledge, gratifying most fully the contemplative instinct, tends also to promote the deeper sort of gratification of the other great instinct, that for practical welfare. This end is the most inclusive singe end we know.[2]

More recently Sheldon has compared philosophy to a world map telling us ". . . where we are and how to get where we want to go . . . It satisfies our inborn craving for knowledge, just for its own sake, and at the same time gives a control of our conduct, a plan of life based on knowledge of our great environment. It is, in fact, indispensable for intelligent conduct of life, even as a man must be able to poise on his feet before he can walk forward."[3]

---

2 Wilmon H. Sheldon, *Strife of Systems and Productive Duality* (Cambridge, 1918), p. 18.
3 Wilmon H. Sheldon, *Process and Polarity* (New York, 1944), p. 10.

By demanding that philosophy draw up a world map to guide our conduct Sheldon does not suppose that such a map can be framed once and for all. "No metaphysic can compass the whole wealth of being; enough if it provides a map that works to a high degree" (*God*, 112). Like S. C. Pepper and H. N. Lee, Sheldon views metaphysics as hypothetical, accentuating the relevance of practical result for verification.

However noble philosophy's theoretical and practical intentions, everywhere its enemies proclaim its futility. Few would, of course, dispute the obvious progress of philosophy, evident in the increased scope and logical exactness of systems since the days of Thales. But despite the progress manifest in the articulation of special systems of philosophy, the simple fact remains that there is little or no agreement among these systems on basic issues. "There is no stock of funded truth in philosophy. Unlike science, unlike practice, it has no consensus of experts" (*Strife*, 26). "After a history of almost unexampled length, philosophy has less of positive information and more of controversy to show, than any other human discipline" (34).

Ancient nihilism bared a triad of negations: nothing is, and if anything is, it cannot be known, and if anything can be known, it cannot be communicated. This nakedness is now clothed in the cautious sophistications of modernist modes of thinking and talking. Clearly, the assertion that nothing is, is directly self-refuted by the consideration that the asserter and the assertion *are*, while the assertion that whatever is cannot be known is self-refuted by the consideration that the asserter at least claims to know what he asserts. Yet the charge persists in diminutive form that whatever is and is known is incommunicable — at least in philosophy. To say so much is presumably to take a post-philosophical stand. Today philosophy is derogated as literal nonsense, and the philosopher is advised to correct his grammar, to mend his language to conform to the rules of precise symbolic techniques or to adjust it to viable ordinary usages. But the derogation turns out to be a species of philosophy, launched in the light of criteria of truth and meaning entailing philosophical conceptions of the nature of language, the nature of the speech community, the nature of the objects of discourse. The indictment of philosophy is prosecuted at a deeper level. The psychology of the philosopher is brought under suspicion. Philosophy, accordingly, is not discourse representing the meanings the philosopher thinks he has in mind. Rather it is expressive of deepseated mental and emotional maladies, symptomatic of unknown psychical diseases. The philosopher is a patient to be cured. When linguistic analysis fails to eradicate his compulsive cramp to do philosophy, psychoanalysis is advised. Still the attack presupposes a tacit philosophy of mind and of culture, and in proposing therapy it is committed to basic philosophical valuations. A minimal philosophy must be conceded, so that next the gravamen against philosophy pivots on an existential exclusion. The philosopher is viewed as a concrete individual imprisoned in his own space and his own time, while philosophy, ostensibly an objective formulation of unchanging truth, is exposed as a flimsy sham to cover the ill-concealed particularities of the philosopher's *milieu*. Again the charge against philosophy stems from a philosophy of society and of existence, a philosophy which paradoxically rules out the existence of the philosopher, and his passionate engagement in the impartial truth.

These tilts with the arguments against philosophy disclose in effect that the arguments are themselves philosophical. Unfortunately this strategy, while increasing the girth of philosophy, neither ennobles it nor renders it less futile. In fact, upon entry into the arena of philosophy even the friends of philosophy encounter a myriad of philosophies, each claiming the truth as its own and refuting its opponents' claims. Interminable disagreements, insoluble wrangles over fundamental issues, testify to the futility of philosophy. Thus philosophy, essentially the activity of making a world map whereby not theory alone but the guidance of human conduct are aided, is perpetually entangled in controversy and disagreement. Its unity is shattered by a multiplicity of vying opposites. On every issue endless wrangling persists. No wonder its nobility seemingly tumbles over into futility.

Unless the nobility of philosophy is to be lost without hope of restitution, it behooves the philosopher to take account of the incessant discord of systems and to proffer principles of reconciliation. Undoubtedly to execute this irenic intention, inescapable for the philosopher, it is necessary to study the divergent philosophies as comprising a special subject matter. Such a study ought, however, to be more than a history of philosophy, because it must inquire into the nature, the logic and the interrelations of the philosophies. This theory of philosophies is, in short, the philosophy of philosophy. When pressed to its farthest implications, it logically involves interpretations of the nature of being explanatory of the variety of systems and so discloses otherwise neglected ontological categories. For over half a century Wilmon H. Sheldon, whose irenic intention its perhaps unequalled in the history of philosophy,

has sought not only to remedy the disease of philosophy, the problem
of disagreement, but also to establish a stable yet flexible philosophical eclecticism. Content neither with the adoption of a partisan
standpoint nor with the elaboration of a new one devoid of reference
to the historical development of philosophy nor with the examination of philosophies as autonomous hypotheses which cannot be
welded into a coherent whole, Sheldon stands out as America's foremost philosopher of philosophy.

II

"All (philosophical systems) revolve in a circle about that
centre which our original problem urges us to penetrate; if we do
not like one place on that circle we may choose another, but the
whole spectacle of human philosophic endeavor offers in the main
nothing more than this ceaseless revolution" (Strife, 412). What
Sheldon calls "the strife of systems"—this is the disease of philosophy. Unconcern in the face of "the strife of systems" may, of course,
be feigned. Philosophy, it may be held, shares with poetry the
virtue of disagreement, as poet and philosopher alike individually
represent or express particular facets of a reality rich beyond the
measure of any theory of concepts or organization of words. If so,
the philosopher, like the poet, would be constrained to the limits of
a single perspective, reality forever eluding any total chart he might
frame. No doubt, differences in personality may account for differences in philosophic system. Temperament molds metaphysics,
but to recognize it does not quell the strife. Besides impugning any
claim to truth a philosophy might make in opposition to the counter-
claims and denials of opposing philosophies, the psychological theory
sanctions an inconsistency that is ruinous of any hopes for system,
since it allows that " 'the world is x' with one man, and 'the world
is not x' with another" (32). Nor can the strife of systems be minimized by regarding philosophy as "inchoate science," as "the residue
of unverified theories" which, when established, constitute science.
For philosophy "contains problems which do not seem open to scientific treatment. The relation between science and faith, the estima-
tion of artistic judgment, of the validity of reasoning and immediate
experience—these are problems of a different sort" (34).

Under the circumstances many philosophers—in fact, most—
have seized the opportunity to take up one partisan position and to
bristle hostile refutations at the others. To Sheldon this would not
do. "If all schools of thought but one are fundamentally in error—
as nearly every body thinks in every age—", he remarked, "would

it not be a miracle that one should escape the common lot? But the most convincing evidence that it is not so is that today, when philosophic interest is livelier than ever before, when discussion is if not keener, at least more widespread than in any preceding age, the refutations are but increased, and the fundamental differences emphasized" (33). Hence rather than be partisan Sheldon has sought to pacify the strife by reconciling and synthesizing the systems. His enterprise has been sustained by three convictions: that since philosophy exhibits an array of recurrent types of system, these systems must in some sense be true; that their contradictions can be mollified; and that the source of the strife, if detected, bears forth an unknown yet fundamental character of reality.

Through the years Sheldon has remained constant in these convictions, though he has revised the lists of rival philosophic types. In *The Strife of Systems and Productive Duality*, an early (1918) and neglected essay which every student who wants to understand the nature and dialectic of philosophic systems ought to master, Sheldon explores a logical progression of systems covering "subjectivism" and its counterpart "objectivism," their opposition resolved in the "solvent" of "pure experience," transcended by "great subjectivism" locked in conflict with "great objectivism," logically succeeded by the irrationalist group—"intellectualism, pragmatism, intuitionism"—, set against the "rationalistic synthesis" of Hegel and his British disciples, coupled in opposition to the "practical synthesis—Thomism." For more than three hundred and fifty pages Sheldon takes his reader on an exciting dialectical journey of inimitable precision and insight. Subsequent study of the main currents of philosophy has prompted Sheldon to alter his original list of the recurrent rival types. Process philosophy he recognized as a distinct type in his Mahlon Powell Lectures at the University of Indiana[4] and in his Woodbridge Lectures at Columbia University,[5] while in his crowning work, *God and Polarity*, an achievement of tremendous scope, scholarly detail, logical acumen, and gracious style, he discusses recent trends in existentialism as forms of irrationalism. Also noticeable is a decided shift from epistemological emphases in the earlier work to stress on the ontological content of systems in the later. At any rate, Sheldon writes: ". . . our list of the more fundamental issues is now: idealism or spiritualism vs. materialism, each and both vs. dualism, rationalism vs. irrationalism, Thomism vs. Process, and within idealism, monism and pluralism

---

4 Published as *America's Progressive Philosophy* (New Haven, 1942).
5 Published as *Process and Polarity*.

or personalism" (God, 157) ; or as these topics are organized to shape up the chapters of God and Polarity: ". . . idealism, monistic or pluralistic, materialism, Thomism, Process, and irrationalism" (158).

According to Sheldon, each type of system is designed to meet a different human craving and hence is unavoidable. Monistic idealism insists on the objective rationality of the universe and its identification with value, personalist idealism focuses on the irreducible reality of finite conscious minds, and materialism vaunts the values of the body and corporeal nature. Synthesizing materialism and idealism, the Thomistic philosophy reconciles their values in an eternal hierarchy of substances rising from a base of corporeal elements, reaching through a variegated domain of hylomorphic creatures, and culminating in a heaven of purely spiritual beings dominated by an absolute Creator. Contrasted with the rigid vertical structure of Thomism stands Process Philosophy, still vague in detail, yet definite enough in outline, to disclose that it "takes time seriously," adjusting the static structure of the Thomistic levels of being to the horizontal dynamics of temporal development. Revolting against the excessive rationalism of the other types, existentialist irrationalism rightly calls attention to the category of crisis and the primacy of the human condition in the construction of theory. Thus each philosophy is right in emphasizing the values for which it stands. In so far as it is positive, it is true. Its truth is verified in the living experience of its adherents. Furthermore, Sheldon's interpretation of what verification signifies in philosophy is extremely open-minded. For example, on the verification of Brahman-Atman or Nirvana he remarks: ". . . the experiment was performed by many men, through many centuries, and it succeeded. The testimony of so many cannot but be accepted. But it is not just verbal testimony; it is witnessed behavior" (217).

Each philosophy, therefore, is true in what it affirms, false in what it denies of the others. What causes the incessant contradiction? Doubtlessly part of the strife can be imputed to the temperament of philosophers, their vanities and hostilities. Certainly, "the weakness of man's reasoning" as well as "the strength of his loves" interplay to produce "the variety and opposition of philosophic types" (140). And in one sense "the root of the trouble" is "exclusive devotion to intellect or reason" (8). Unlike his placidly tolerant Eastern counterpart who lives amicably in the company of contrasting philosophies and who justifies his standpoint by the way of direct experience, the polemical Western philosopher relies upon reason alone as the guide to truth. Claiming rigorous demonstration for his own system, he apprehends that his rivals' systems lack such demonstration, and so, overlooking that in this respect his own system is in equal straits, he concludes wrongly that his rivals' theses are false. In *The Strife of Systems and Productive Duality* Sheldon ascribes the interminable controversy of philosophies to a logical predicament which constrains them to contradict each other. In *God and Polarity* he traces the genesis of disagreement back even beyond this logical germ to its source in the philosophers' loves and valuations, to their passionate worship of some exclusive category and their consequent blindness to others. He observes: "Disagreement is due to disvaluing . . ." (140).

Sheldon's diagnosis of the logical germ in *The Strife of Systems and Productive Duality*, unstressed in the recent work, deserves reexamination. For if the analysis is correct, it shows that contradiction is inevitable so long as philosophy is a congeries of systems each short of the total truth, while it offers a theory of negation which eradicates these contradictions by literally transforming them into non-contradictions. The constraint to contradict is attributed to the fact that ". . . our thought cannot help being governed by two principles which appear to contradict each other" (*Strife*, 422). The "disease-germ which has poisoned all human philosophy" is nothing less than "the apparent hostility" between the principle of internal relations and the principle of external relations (423). "The doctrine of externality says that there are entities which are the same in all environments, independent of the relations in which they enter, and ultimate" (415). Every system in opting for the ultimacy of its category — matter, mind, etc.—acknowledges the principle of externality in respect to this category. Its category is believed to be ultimate—nay, it is experienced as ultimate. Yet each system is the attempt to *understand* that which lies on the other side of its favored category, and it can do so only by appeal to the principle of internality. It must, in other words, reduce all other categories to its own or to some relation of its own: materialism, for instance, defines mind as relations of matter. "It is as if each category said to its counterpart, 'I am ultimate and you are not, for you are only a relation in me' . . ." "While each system saves itself from internal contradiction by applying externality to its own basal category, internality to its counter-category; as a whole, the field offers the spectacle of a contest between these two antagonists" (417). Though perhaps the strife could be eliminated by sacrificing one or the other principle, Sheldon is emphatic that both are necessary for thought. Without internality the universe could never be understood as a single whole. Without externality no category could be singled out as an object of belief or experience. Meanwhile, beyond the strife of systems lies a real world at peace. "Somehow the real world itself has harmonized these antagonisms: if it did not, it would be instantly annulled. As a man who contradicts himself takes away our belief in what he has just said, so a reality which was inconsistent would remove what it put down— and we should have no experience" (453).

III

"The deepest trait of reality, in short, that which makes it the moving, productive thing it is, is just this marriage of two principles whose apparent hostility has constituted the continual frustration of man's effort to map the universe" (*Strife*, iv). "Polarity" is Sheldon's term for this positive principle. By means of it he seeks to pacify the strife of systems and to integrate their truths within an embracing synthesis. The discovery of polarity (or duality) is the fruit of an exacting search for the root of the trouble with philosophy. The putative antagonism between internality and externality, Sheldon found, ultimately stems from the supposition that "sameness and difference are deemed incompatible" (455). But against this allegation of incompatibility Sheldon has contended that it ". . . is the purest dogma, a fulmination out of the darkness, justified by no utility or self-evidence" (456). Experience continually reveals that things and qualities are the same yet different, that sameness and difference are not mutually exclusive. Red is the same as blue in that it is a color, different in that it is patently not blue. The apple is red and sweet, yet being red is different from being sweet. Purple is simply one color, yet by containment it is red and blue. Insisting upon the compatibility of sameness and difference Sheldon nevertheless refrains from ". . . asserting that any quality can combine with *any* other quality . . . (He is) only concerned to deny that in a logical point of view *none* can . . ." (468).

What causes the dialectician, in behalf of some favored category upon which he builds his system, to refute all other systems based on other categories different from his own? Sheldon's considered answer is:

> . . . that little word "not" which contains just enough of ambiguity to misdirect the intelligence. For we use that one word to mean now the relation of otherness between terms, now the denial of a suggested judgment. Had the peoples who gradually formed the English language been, *per impossible,* exact logicians, they would

have used two different words for these distinct meanings. When
I say, Caesar did not kill Brutus, I contradict the suggestion that
he did so; when I say red is not blue, I may mean only to signify
the duality of these two. The former proposition is equivalent to
"it is not true that Caesar killed Brutus"; the latter proposition need
not be meant as a denial of anything, but rather as an affirmation
of the relation *otherness* or *duality*. *Otherness* is a very different
concept from *opposition* or denial; unfortunately the negative of
human language has not indicated the difference" (471-472).

Accordingly, the categories represented by the rival philosophies
are real, and in so far as they are affirmed by these philosophies,
the philosophies are true. Just as the idealist is right when he says
mind is reality, so is the materialist right when he asserts that
matter is reality. Error arises when either, perceiving the difference between mind and matter, fallaciously infers the denial of the
other or wrongly attempts to reduce the other's truth to a relation
of his own favored category. The denials consequently derive from
a fundamental misinterpretation of the nature of negation, since
they fail to grasp that in the philosophical context negation means
"... *otherness* rather than *removal* of what is negated" (524-525).

Logical considerations illuminate the compatibility of sameness
and difference, a compatibility which allows the coexistence of polar
categories. That this polarity is more than a mere possibility but
constitutes instead the vital and basic principle of reality, is not
only borne out in the logical juxtaposition of the historic philosophic
systems; it is also confirmed by direct ontological investigations.
Sheldon's sketch of the way the divergent systems exhibit polarity
is unfolded with comprehensive scope and probing detail in *God and
Polarity*. After listing the systems: idealism, materialism, Thomism,
process philosophy, and irrationalism, Sheldon remarks: "Note that
the systems have paired themselves off. Each has its *specific* opposite, its bitterest foe, denying just what seems to it the principle or
principles of maximum value to man." "Note also: this very procedure, forced on us as it were by the nature of the oppositions,
suggests the way of reconciliation. If each type has its specific
counterpart, should not the harmony be a polar union, a lawful
wedlock of counterparts?" (*God*, 158). Monistic idealism and per-
sonalistic idealism yield respectively the polar categories of the
structured whole and of the free individual, both idealisms witness-
ing to the reality of mind, which stands in polar contrast with the
category of matter contributed by materialism. Then Thomism,
proffering a synthesis of mind and matter in the doctrine of hylo-
morphism, provides the levels of being linked by analogy and cre-

ated by God, while its polar synthesis, Process Philosophy, affords novelty and adventure by imparting temporal dynamism to the static fixity of the Thomist hierarchy. The reconciliation of Thomism and Process Philosophy, moreover, is suggested in light of the special gift of existentialist irrationalism—the category of crisis, expressive of the conjunct of self-growth and self-liquidation experienced by an individual in its transition from one level to the next in the dynamic hierarchy of being.

The principle of polarity, the doctrine of counterpart pairs, is therefore underscored in the array of systems, and serves, when its logic is understood, as the key to their reconciliation. For polarity is an ultimate ontological feature, a neglected basic characteristic of reality. What is polarity? Pervading all that is, polarity is thus defined: ". . . one or another phase, aspect, relation, event or entity and its counterpart; the two opposite as it were in direction, in way of acting, yet each capable of fruitful cooperation with the other, also of opposing, denying or frustrating it, having thus a degree of independence and a being of its own, and between the two a trend or lure of cooperation in which one of the partners takes the initiative and the other responds, yet each freely; the relation has a certain asymmetry" (674). To the individual terms of this definition, or description, of polarity Sheldon gives special attention. The polar pairs are counterpart couples which, although each is in some measure independent of the other, are linked together like the left hand and the right hand of the same human organism. Ideally the couples cooperate, so that the loss of a member of the pair frustrates their combined function. Because cooperation requires difference of direction for each of the polar couple, they are opposite but not opposed, as for example man and woman are in the ideal marriage. Nevertheless, their cooperation is not logically compelled; it is free. Ideally each member cooperates with the other by virtue of its own autonomous efficacy. When opposition does occur with compulsion following after or conversely, evil results. Further, the couple are related asymmetrically; one member, though needing the other, is still superior in power and initiative. By virtue of this asymmetry stalemate is minimized. Should opposition or compulsion or evil occur, the asymmetry triggers off the onward push of the situation to a higher resolution.

Here process enters the picture. In *America's Progressive Philosophy* Sheldon singled out Process Philosophy for its promise and its progressiveness. He there regarded it as intrinsically superior to the other philosophic types since it is ". . . able to grant them all an equal truth, and a truth as good as its own . . ." (*America's*, 3), and since it ". . . has a welcoming and generous attitude, calling for cooperation to replace competition, and for new and venturesome points of view of the inexhaustible wealth of reality about us" (4). In *Process and Polarity* Sheldon systematically explores the significance of process and accommodates it to his long-held philosophy of polarity. When the opposition of polarities mirrored in the philosophic systems is actualized in the concrete conflicts that mark natural and social realities, process assuages the evil that these clashes generate. "For process is nature's great remedy, the healing potion supplementing the imperfections that mar the polar order, the hope and lure of the future and the basis of a working confidence in progress. It is the office of the process-principle to remove the clash and conflict between the polar opposites, restoring their energy for the free variation they should have unmolested" (*Process*, 118).

Thus Sheldon's philosophy of philosophy entails a philosophy of polarity. The argument, then, is conducted beyond the consideration of philosophies to an objective metaphysical investigation of the polarities that throng the universe. The structural as well as the dynamic categories of ontology pair off respectively into essence-existence, act-potency, substance-accident dyads and into the dyad of efficient-final causality. Similarly nature, pervaded as it is by the form-matter polarity, manifests ubiquitous counterpart pairs in cell life, in the human individual, in interpersonal relations, in social organization, in historical movement. And the climaxing note of Sheldon's argument focuses specifically on "the basic human polarity, man-woman" (*God*, 708). As the highest exemplar of free, cooperative polarity Sheldon cites the holy matrimony of man and woman.[6]

IV

"Synthesis is the life-blood of philosophy . . ." (*God*, 424). Inimical to the synoptic vision which philosophy, unlike the special sciences, promises, however, are the partisan systems, for they select certain categories and ignore the others, and then worst of all turn their ignorance into denial. The fulfillment of the promise of

---

6 Sheldon is unabashed about the sexual overtones of his doctrine of polarity, gladly avowing that ". . . *man's mind* works by pairs. A Freudian would ascribe this to a subconscious sex obsession. Perhaps he would be right; if so it is a happy one" (*God*, 673). Sheldon's theorizing about sex receives its fullest treatment in *Sex and Salvation* (New York, 1955), a delightful, controversial philosophy of woman. As its Preface declares, ". . . its central thesis lies in a spiritual interpretation of woman's anatomy and physiology and the mind that goes with them, as revealed in comparison with man's. It assumes the basic truth of Christianity."

philosophy, therefore, requires synthesis. But true synthesis, as Sheldon contends, must respect the integrity of the categories it embraces, since the very survival of the types of systems bears witness to the autonomy of the categories upon which they are built. Nor can the true synthesis merely collect all these categories together in a heap, because such an aggregate is hardly true of the coherence which philosophy expects from an ordered reality. Hence the synthesis must be systematic without annulling the independence of its components. Consequently, Sheldon explicitly judges monistic idealism unsatisfactory in as much as it derogates the reality of all beings except the Whole; its failure is that ". . . it includes by degrading to the status of abstraction and error . . . Genuine inclusion treats its guests not as inferiors but as equals and friends" (345). On the other hand, of Thomism he remarks: "Without question it is the fullest synthesis as yet offered to thinking man" (441), although its shortcoming is an excessive intellectualism which subordinates the active phase of man's life to the contemplative. This shortcoming Sheldon rectifies by making action ". . . coordinate with intellect and intelligibility . . ." (502). Further, in *God and Polarity* Sheldon has Thomism assimilate the positive dynamism of Process Philosophy. "The Thomist *can* admit the 'new logic', as we call it, of self-alteration or self-expansion, and to a high degree has already taught it. The opposition between the ultimate reality of substance, and that of becoming, is needless." "Substance is essentially dynamic . . ." (500).

Sheldon's philosophy of philosophy culminates in a philosophic synthesis of Thomism and Process Philosophy, with Thomism accorded the position of priority. At this juncture the examination of Sheldon's thought shifts from the consideration of his comparative study of philosophic systems to an evaluation of his philosophy as an adequate theory of reality. Now Sheldon maintains that the decision between warring philosophies depends on an explicit criterion of reality. He also recommends the formulation of this criterion through the involvement of the whole man. Stimulated by decisions which spring from practical and theoretic needs, thought is not the sole approach to reality, nor can it reach reality alone, for "thought deals with possibles" (100) while "action is the door-opener to reality" (38). Internally, action is effort, and effort ". . . is in fact the very essence of the self, the wellspring of its worth and its growth. A man is what he does" (31). Exerted outwardly, effort is action which encounters resistance by objects: ". . . what makes the object something by itself . . . is its power. As John

Dewey has said, the object is that which *objects*" (36). It is action, then, which connects power and reality: ". . . power is a sign of, or rather identically is, a being or thing or event 'on its own,' existing in its own right, real. It is the *esse* of every *ens*" (37). Precluded from ascertaining the power, the *that* of reality, which is open only to action, thought is nevertheless indispensable since it uncovers the *what* of reality. "The full sense of reality is not furnished by thought alone, nor by action alone; only by both together. Action without intellect is blind, intellect without action past or present is empty" (505). Thought discloses order, action power, and the internal polarity of thought and action as well as the external polarity of order and power conjoin in man's affective phase and its objective counterpart — value. The polar opposites, furthermore, are lured together. "*Lures* is the word, not *compels*. It works as if by drawing, not pushing, by urging, not forcing, shall we say by love rather than force" (14). Hence "reality . . . is power, order and value" (123).

Few would deny that a criterion of reality is necessarily prerequisite to any decision between rival philosophies. But to provide such a criterion is just to advance another philosophy fraught with all the controversy it is intended to remove. Nor would many dispute that reality is properly grasped by the whole person as composite thinker, lover, agent, though the ordering of these aspects of the whole person comprises a philosophy of man equally subject to controversy. Experiment in the scientific sense is not helpful, since it is admissible alike by all the major rival philosophies and serves solely the restricted purposes of science. And appeal to experience does not settle the question, because the significance of experience is itself a philosophical thesis, variable with the variety of systems. At its best, then, Sheldon's criterion of reality, unable to do the job for which he designed it, still bespeaks a broadly tolerant philosophy able to contain others or to accommodate itself to them. In this sense it is a brilliant articulation of *philosophia perennis* marked by a certain novelty of interpretation of central traditional theses.

The solvency of Sheldon's synthesis, untestable by its criterion of reality because it includes this criterion, pivots on consistency. For unless the synthesis is logically sound, it will be marred by irreparable fractures. Central to this synthesis is polarity, manifest in Sheldon's preoccupation with it since the earliest work. The very title of the first book mentions "productive duality." There he wrote: "Duality and unity are not of quite *equal* rank: the account we shall give of the universe will describe it primarily as a duality"

(*Strife*, 475). Later, of course, he confessed the inadequacy of polarity, cautioning: ". . . let no metaphysician claim to have unearthed the one absolute all-consuming principle that dominates every reality, actual or possible. For polarity we can make no such claim. We affirm only that it is a very widespread phenomenon, fertile and suggestive and indeed, being polarity, indicating a counter-principle" (*Process*, 106). Simultaneously he singled out process as the counter-principle. At present, however, Sheldon accords this role to the ". . . levels of being, pictured vertically by the Thomist as pointing up to the perfect Creator, more horizontally as a long ramp by Process, leading to a fuller life for humanity" (*God*, 673). Polarity, then, itself is a counterpart principle, and Sheldon's synthesis, straining to weld together Thomism and Process Philosophy in a doctrine of levels of being polar to polarity itself, suffers severe fractures. These fractures are evident in the conceptions of God and of possibility.

God is not a polar counter-principle of the created universe. "The relation of Creator to creature is not polar. Creation is a free act; there is in God no need of the creatures in the sense that He requires their cooperation for His own fullest being" (678). God is like "the lover of great music" who ". . . . needs no one to share his joys, though he welcomes it gladly and freely." "The note is this: *I need you not, but I gladly welcome you*" (286). "So God loves His creatures, not at all because they are His relatives and He likes company, still less because He needs them. He loves them because they need help . . ." (288). Apparently, then, Sheldon's motive for refusing to regard the relation between God and the World as polar springs from his fear that otherwise the perfection of God Who does not need the World would be impugned. Accordingly, the implication would seem to follow that, despite Sheldon's perpetual insistence that polarity is "free", somehow it is infected with too much compulsion to be extended over the bridge between God and the World. Besides, by discounting polarity in this case, Sheldon rejects the only principle at hand with which to relate God and the World, so that this relation subsequently eludes the possibility of rational understanding. Paradoxically, when Sheldon tries to discuss this relation, he perforce falls back on the language of polarity, going so far as to concentrate on a given human experience to disclose a positive principle which, though unattainable by reason, presumably illuminates the relation between God and the World— "the principle, shall we call it, of a *loving* duality: self-sufficiency that welcomes the other for the other's own sake" (294). Naturally

Sheldon disregards the contradictions resultant from this procedure, attributing them to a narrow rationalism and promising that ". . . what is contradictory for a lower may be consistent in a higher dimension" (291). Thus, by confining polarity to the World while affirming God as its Creator, Sheldon fails to set up a relationship between God and the World consistent with his philosophy, and consequently, it may be said, he surrenders his philosophy to his theology.

Whereas Sheldon's philosophy of polarity buckles to the religious demands of his synthesis, his Thomism in turn is forcibly fitted to a novelty proposed in this synthesis—the doctrine of possibility. In nature evidence for a realm of possibles appears in the operancy of chance and law: ". . . today we can affirm that not only anything within certain limits *may* happen, but everything within those limits *will* happen, yet in no predetermined order. Nature leaves to the individual events a chance wandering among the alternatives yet holds them in the long leash of equality between them all" (513). Nor are these possibilities reduced to mere adjectives of substances. Transcending all past and present actualities, *"possibles make progress possible"* (517). The very advance of science ". . . has been made by intellect breaking away from things as they are given to sense: not merely penetrating by abstraction into their essences, but by venturing to ask what *might* be the case" (518). And mathematics and art both reach to ". . . unrealized possibles, even possibles as yet unrealizable in our world" (519). Hence "this field (of pure possibles), infinite in extent, implicit indeed in the principle of plenitude long known of old, has emerged into the clear vision of modern man; it gibes well with the recent process-metaphysic which ever soars toward the novel . . ." (509). Now Sheldon claims that ". . . though it (Thomism) may not *admit* that realm, it *permits* the same" (508), next declares: ". . . the neo-Thomist *could* admit this infinite realm of possibles, possibles not dependent on this world, dependent on nothing at all, not even on the Divine Will, which is concerned only with creating those of them which He chooses" (509).

At this point difficulties crop out in Sheldon's synthesis. On the one hand, "God is the realization of all possibles in one act of being." On the other hand, as the next line states, ". . . there are possibles not included in this all-inclusive act, namely, some of them as *excluding* others" (519). The excluding possibles in effect comprise the created World, and are not included in God, because *"God includes no exclusions"* (520). In addition to these possibles located in our world, it is suggested that there are other possibles, ". . . possibles not yet present in our arena, yet luring us by their ideal beauty to progress to new and better ways of living" (520). Where is this realm of possibles? "It is not in the created universe, it is not in God's mind: the answer is, it has no locus . . . it is an ultimate category or transcendental, a pure potency not resident in any act, all alone by itself, subject to God's will as regards becoming actual" (520). So both the created world and the realm of possibility are not included in God, the former as an inferior dependent marked by exclusions, the latter as an externality objective to God in the manner of the Platonic Heaven of Forms to the Demiurge. Coeval with possibility, the God of Sheldon's synthesis seems to lose the primacy of position accorded God in Thomism. Immediately, however, Sheldon retrenches on this implication, asserting that ". . . with regard to the ultimate possibles: they are all present in the Divine being, none excluded. If some were excluded, a cause would be needed. But none are so, hence no cause is needed: God's existence is self-explaining, necessary in itself" (527). As Sheldon puts it, this is ". . . the old ontological proof, but in objective terms" (527). In accord with this interpretation it would seem that the inclusion of all possibles in God — those constituting the created World as well as the realm of possibility beyond — undermines the sharp demarcation between God and the World. Once again Sheldon's synthesis totters near collapse. Basic theological convictions consistent with Thomism conflict with fundamental philosophical theses borrowed from the process-metaphysics. So long as Sheldon adheres to the doctrine of possibility in its present form, his synthesis conflicts with the theology of Thomism, a theology which in important respects he accepts.

Of course, these objections against Sheldon's philosophy may perhaps be brushed aside by deeper study of this superb metaphysician's subtle and comprehensive system. After all, controversy is always a danger in philosophy, especially for a thinker who intrepidly dares to reconcile philosophies. This does not, however, soften the conviction that Sheldon has failed to quell the strife of systems. But in the attempt to do so, he has, first, formulated a most impressive philosophy of philosophy, doubtlessly the major one of our time; second, he has correctly accepted the challenge which must be faced by every metaphysician who wishes to master the future course of philosophy — the reconciliation of Thomism and Process Philosophy; and third, despite difficulties issuing from its broad eclecticism, his philosophy with its novel emphasis on polarity is a brilliant synthesis, an original articulation of *philoso-phia perennis*. If Sheldon has failed to resolve the strife of systems, he has at least shown that it is inevitable as long as men who want the total truth come up with systems which are less than that. And he has indicated how the most noxious and disruptive features of this strife may be averted, by practicing a broad tolerance of op-posing philosophies on the grounds that they, too, testify to truths which, though partial, must be contained in the final account. Reconciling synthesis is the byword. Could any philosophy reason-ably expect it to be otherwise? Sorry, indeed, would be agree-ment if it were achieved with less than the total truth or a close approximation thereof, and certainly, when this truth is reached, agreement will follow after — among philosophers at least. In the meantime, every approach to the total truth must be recognized as a tentative system subject to revision, and the nobility of philosophy must be seen as a virtue of the philosophers who, in face of the arduousness of their enterprise and the oft-proclaimed futility of their vocation, persist in the quest.

IS THE STUDY OF AESTHETICS A PHILOSOPHIC ENTERPRISE?

Louise Nisbet Roberts

WILLIAM JAMES has pointed out that as soon as a philosophical problem is definitely solved, it is automatically claimed as a part of science, with the result that philosophy is left "holding the bag," and is restricted to questions that have not yet been answered. That there are many unanswered questions in aesthetics is not to be denied. The field of aesthetics is notoriously confused and many questions in aesthetics are nonsensical in more than the positivist sense. Despite this situation, however, there are some claims that aesthetics is a science.

One may find "aesthetics" defined as a science in such diverse reference works as *The Catholic Encyclopedia, The Columbia Encyclopedia,* and Webster's *New International Dictionary.* Such definitions do very little to clarify the state of confusion. In some cases one may recognize from the context that the term "science" is given a very general meaning, so broad in fact that all philosophy might be regarded as science and hence aesthetics would be no different from other fields in this respect. In other references, however, "science" is used in a narrower sense so as to suggest that aesthetics is ready to follow other branches of learning such as astronomy, physics, and psychology in declaring itself relatively independent of philosophy, the mother discipline. One may ask, therefore, whether aesthetics is science or philosophy. If it is a combination of the two, in what respect does it partake of each? If it is a branch of philosophy which may become a science, how might such an emancipation be accomplished?

One of the major "idols of the market place" in this problem is the term "aesthetics" itself. Although less emotive than the term "beauty" with which it has been related as theory to subject-matter, it has acquired emotive force in about two hundred years of popular usage. Furthermore, the term has undergone rather drastic semantical shifts of reference. In the works of Kant it is shifted from the more etymological meaning "concerned with perception" and made to refer to an investigation of the judgment of taste. Both of these meanings have become involved in various other interpretations. Aesthetics has been developed as an analysis of "aesthetic experience" and as a "theory of taste." It has often been regarded as a branch of axiology. In some cases focus has been placed upon immediate experience or experience of value, and in other cases emphasis has been upon judgments or evaluations. Furthermore, although Kant's investigation of the judgment of taste includes consideration of natural as well as artificial objects, the term "aesthetics" has undergone another shift (attributable to Hegel) to refer to investigations of art exclusively, and is often used as a synonym for philosophy of art. In recent years it has been suggested that the meaning be restricted to apply to the "science of art." Although there may well be further ambiguities, these examples are sufficient to indicate that arguments concerning the relationship between aesthetics and science depend *inter alia* upon one's conception of aesthetics and, in some cases, difference of opinion may rest upon quibbles regarding the meaning of the term.

Different conceptions of aesthetics need not be attributed entirely to confused thought or careless use of language, however. They may be accounted for in terms of a variety of related problems some of which lend themselves more or less easily to scientific solutions. An investigation of some of these problems should reveal something of their further dependence upon philosophy and may suggest possibilities for future exploration.

The notion of aesthetics as a "science of art" has developed as an attempt to solve problems revealed by increased awareness of the multiplicity of the arts and their tremendous significance in various cultures, including our own. This development may, in turn, be attributed to advancement within such fields as psychology, anthropology, archeology and art history. It has been aided by technological developments such as art photography and sound reproduction. Extensive empirical study of the particular phenomena of art and art creation is now not only a possibility but a challenge. Factual description of the arts and of value judgments as well can now be made. An inductive procedure from observation to tentative hypotheses and on to the application of these hypotheses to particular cases can now be followed with respect to the arts. Such a view is championed by Thomas Munroe, and through his writings one may glimpse an exciting vision of this "new science."[1] Although Munroe admits the difficulties of quantitative measurement and laboratory procedure within the field, he argues that increased objectivity and logical organization is possible. One gathers that the empirical science is "on the way." Is this empirical development the task of philosophy?

Whether or not aesthetics as a "science of art" is to be developed by philosophers may depend in part upon academic conventions and also upon the major fields of individuals contributing to such a development. It involves what might be regarded as an interdisciplinary approach. Such generality could place it under the aegis of philosophy. But generality is a matter of degree and it is to be remembered that pioneers in psychology were drawn from such fields as physiology and medicine as well as philosophy. Also, the extent to which empirical investigations of art can be made scientific remains to be seen. Difficulties of quantitative measurement and laboratory procedure have been indicated. These investigations may be destined to occupy that broad limbo between exact science and philosophy which is already well populated by other carefully disciplined bodies of knowledge which focus upon the observation of particulars.

Admitting the possibility of a scientific aesthetics in the above sense, it can be seen that certain problems remain which are more recalcitrant to solution by scientific method. In order to be meaningful, any investigation of empirical data demands some sort of theoretic framework, however tentative. It requires both selection and organization; art is no exception. At present there would be some difficulty in isolating the phenomena for such a study. One is faced with the question of what is art. This is a traditional problem of aesthetics as philosophy of art. The difficulty may be attributed, in part at least, to the ambiguity of the term "art." This is one of the many areas in which there is need for philosophy as linguistic analysis. Art has never been given a generally acceptable definition; but underlying confusions of usage and differences of definition are various theories of art. Art is a matter for theoretic speculation.

---

1 Thomas Munroe, *Scientific Method in Aesthetics*, N. Y., 1928; "Aesthetics as Science: its Development in America," *The Journal of Aesthetics and Art Criticism*, 1951, Vol. IX, pp. 161-207; "Aesthetics," *The Dictionary of Philosophy*, ed. Dagobert D. Runes, N. Y., n.d., p. 6.

Not only is recognition of differences of interpretation required before empirical investigation can be given orientation, but a stand must be made, however tentatively, with respect to these interpretations . Such a stand can be made without defense or elaboration, but this is an avoidance of the problem rather than a solution. To take a stand as to the nature of art and to defend it against opposing views is to develop a philosophy of art. It is within the framework of such a philosophy that questions of form, expression, function, etc. may be approached. This even applies to judgments of taste with respect to art. Before these judgments can be described, some assumptions must be made as to what they are about and how they may be related to judgments of taste in general. There is a further issue as to what constitutes a judgment of taste, but this leads beyond the realm of art.

Selection, therefore, is involved in the organization of an empirical investigation of art and this selection rests upon an intellectual evaluation of theoretical assumptions, the choice of a philosophy of art. This is not the only place, however, where evaluation is involved. It is implicit in most, if not all philosophies of art. The answer to the question whether or not given phenomena constitute art depends upon evaluation of the phenomena in question. If art is defined in the simplest sense as any artifact or product of human activity, this evaluation is relatively simple, but the field of art becomes so broad as to include the entire study of man. As a rule, therefore, the field is narrowed to include only artifacts or human activities which possess certain characteristics. The characteristics provide the criteria according to which evaluations can be made. For example, should "art" be used in the simple sense of "skill," then skill would provide the criterion according to which a phenomenon can be judged as art. As can be observed from this example, these criteria may not merely serve to define the field of art but can also function as criteria for judgments within the field, e.g., the greater the skill the greater the art. It can be argued, therefore, that a study of art involves judgments of value in which certain assumptions are made concerning value criteria and the nature of value *per se*. The study of art can be regarded as an instance of applied axiology.

Value judgments within the field of art suggest further theoretical issues. As has been mentioned above, before these judgments can be described, some assumptions must be made as to what they are about. A description of these judgments rather than others rests upon an interpretation of art as their subject-matter. Furthermore, one is not only faced with the study of art and judgments concerning art but there is the practical problem of making such judgments. Assumptions concerning criteria of art value and the nature of art are implicit within art criticism. Art criticism as well as the study of art may be regarded as applied philosophy of art and axiology. In this case as in others it may be possible to avoid the theoretical problems, but it is hardly desirable. Judgments of art value can be described without any attempt being made to discover the principles of criticism. Art also may be criticized without principle. But a mere description of judgments of art would seem to be of little but historical or sociological interest, and unprincipled art criticism seems hardly desirable. The theoretical issue of interpreting art and discovering the norms of criticism remains.

These judgments of art value can also be regarded as judgments of taste with respect to art. This introduces the question of what is meant by taste, and speculation concerning this problem leads to theories of taste. In the development of these theories, judgments of taste may be restricted to judgments concerning art, but the usual interpretation is much more general. They can be regarded as expressions of preference among values of a certain kind which are to be found in situations involving natural as well as artificial objects. The problem then arises as to the distinctive type of value about which these judgments are made — hence the theory of "aesthetic value" as a branch of axiology. As in the case of art criticism, particular judgments of aesthetic value can be described without any exploration of their foundations beyond description of their social and psychological contexts. Descriptions of this sort could be used for the prediction and possibly the control of such preferences. If so, their value for such fields as advertising and propaganda would be tremendous. But, again one is faced with the question of norms. The description of normative judgments does not solve the problem as to the nature of such norms. In pursuit of a solution, one is lead to an analysis of the experience which prompts these judgments, in this case "aesthetic experience." Speculation develops as to the nature of this experience, characteristics of the value found therein, and the foundations of aesthetic value.

Of course it is possible to take a positivistic position with respect to these problems. It may be held that aesthetics, like ethics, is, as a branch of knowledge, comprehended in the social sciences. One might agree with Ayer that the only information which we can legitimately derive from our aesthetic and moral experiences is information about our mental and physical make-up. But the positivist reduces aesthetics to social science by fiat. Problems which do not lend themselves to solution by scientific method are automatically excluded. This attitude is most frustrating to those possessed of intellectual curiosity, a gad-fly which has been stinging philosophers into action throughout the centuries.

A survey of the various problems mentioned above will reveal that many are extremely abstract and, in some cases, quite removed from the particulars of experience. This presents a danger in speculation as to their solution. As Kant has warned, concepts without percepts are empty. Theorists must return continually to the "cave" of experience. Many years ago a youthful Santayana remarked that writings about beauty fall into two groups: "that group of writings in which philosophers have interpreted aesthetic facts in the light of their metaphysical principles, and made of their theory of taste a corollary or foot-note to their systems; and that group in which artists and critics have ventured into philosophic ground, by generalising somewhat the maxims of the craft or the comments of the sensitive observer." [2] Santayana was a leader among those who have attempted to bridge this gap. Today a similar division may be observed between those who concentrate upon the traditionally philosophic, theoretic, speculative problems of aesthetics and those who focus attention upon direct experience of the arts and the findings of empirical sciences which generalize from particulars. Such division is detrimental to both ventures. There is need for the development of scholars who can combine in one person aesthetic sensitivity, wide experience with aesthetic phenomena, extensive knowledge of the contributions of pertinent sciences, and awareness of philosophic implications. Such scholars are perhaps ideal. A more practical solution may lie in greater cooperation between those who possess some of these qualifications. This cannot be looked upon as a "departmental" undertaking in the narrowly academic sense.

But is aesthetics philosophy or science? With the advance of the sciences it can be expected that an increasing number of the problems of aesthetics will be subjected to investigation by more or less rigorous scientific method. Perhaps, as James suggests, they will eventually be claimed by science. One cannot dogmatically exclude the possibility of even a science of norms. But the field presents many questions to which it is extremely difficult, if not impossible, to establish scientific answers. Aesthetics as an investigation of these problems is, and promises to continue to be, a philosophic enterprise.

---

2 George Santayana, *The Sense of Beauty*, N. Y., 1936, p. 4.

PHILOSOPHY AS COMPARATIVE COSMOLOGY

Robert C. Whittemore

"In putting out these results, four strong impressions dominate my mind: First, that the movement of historical, and philosophical, criticism of detached questions, which on the whole has dominated the last two centuries, has done its work, and requires to be supplemented by a more sustained effort of constructive thought. Secondly, that the true method of philosophical construction is to frame a scheme of ideas, the best that one can, and unflinchingly to explore the interpretation of experience in terms of that scheme. Thirdly, that all constructive thought, on the various special topics of scientific interest, is dominated by some such scheme, unacknowledged, but no less influential in guiding the imagination. The importance of philosophy lies in its sustained effort to make such schemes explicit, and thereby capable of criticism and improvement." [1]

THREE decades have elapsed since Whitehead's attempt to set philosophy to this "sustained effort", and the philosophic scene is more crowded than ever with empiricist, existentialist, and positivist partisans of the "detached question". Where philosophy has not become anti-philosophical it has become ancillary to science. "Constructive thought", for many today, has no other meaning than the activity of the theoreticians of the exact sciences. Should we then accept as fact that Whitehead was wrong? To answer either way is to state the nature of the philosophic enterprise.

The issue is fundamental. If Whitehead's conclusion is rejected, philosophy resolves into an adjunct of physical science or psychology. Empiricism and existentialism can, in themselves, aspire to no higher goals. Alternatively, to assert the autonomy of philosophy is to envisage the task of the philosopher as the framing of the

---

1 Whitehead, A. N. *Process and Reality.* New York: Social Science, 1941. p. x.

universe in some scheme of ideas whereby all experience, personal or scientific, finds its interpretation and explanation. Between philosophy as adjunctive and philosophy as autonomous there would seem to be no settled middle ground. The direction of one's view as to the nature of the philosophic enterprise has been irrevocably determined when one or the other of these alternatives has been chosen.

Because I am convinced that no question is finally resolvable in detachment, that any one question leads only and always to another, and that to another still, until the resolution of one becomes the resolution of all, I choose autonomy. In so doing I do not deny that some questions can be profitably pursued in *relative* detachment. But relative detachment is not absolute detachment, nor could it be for the simple reason that no question, no philosophic problem, is free from presuppositions. Indeed, the very recognition of a question or problem as such seems to involve at least two presuppositions; one, that a question or problem really exists; two, that some approach to it is necessary or possible. Again, to say this is not to repudiate detached questions, rather it is to demand with Whitehead that they "be supplemented by a more sustained effort of constructive thought".

We will be told by the logical empiricist that this is not necessary because the issues involved are at bottom semantic. And the existentialist will remind us of the utter individuality of the experience wherein all questions arise. But the existentialist is prone to forget the self-transcending capacities of the individual, and the empiricist too often tends to overlook that positivism also has its presuppositions. Concerning such adjunctivist views we will have more to say below. Our present problem, having chosen the alternative of autonomy, is to see just what such a choice implies as concerns the doing of philosophy per se.

Autonomous philosophy is that enterprise wherein the philosopher proceeds to the framing of a synoptic and internally coherent categoreal scheme explanatory of all experience, personal or scientific. The important thing to mark concerning this definition is that it is neutral as regards the adoption of any one particular categoreal scheme. To see why this is so, we have but to recall that no philosophy, systematic or otherwise, is free from presuppositions. For if there is any one feature indispensable and common to all viewpoints it is surely the presence, explicit or implicit, of presuppositions. There is not, nor could there hardly be, a point of view totally presuppositionless. The history of speculation discloses no metaphysics, no empiricism, no existentialism, for which some fact or factor — unproven — is not simply presupposed as given. This being so, it would seem that if we are ever to arrive at a sound basis for choice as between competing philosophies, it can only be as the result of analysis, comparison, and justification of those primal assumptions on which the philosophies in question rest.

The difficulty of the task should not be underrated, for while no philosophy lacks presuppositions, few are these philosophers who recognize the foundations of their views, fewer still are those who expose these foundations to critical examination. Philosophers who have not ignored presuppositions have tended to take them as self-evident. In the aura of silence and ignorance characterizing this topic one finds the locus of virtually every philosophic dogma, the crux of most philosophic controversy. In the absence of critical discussion of the nature, type, and validity of presuppositions every conceivable variety of philosophic nonsense has prospered and flourished.

Hence if I would seek to know why I should follow Whitehead rather than Aquinas, prefer Hegel to Aristotle, I must find my criteria in that analysis of presupposition which is the prius of all comparison of cosmologies, and any choice of system. Philosophy today lacks, yet must develop, a methodology for the criticism of its ultimate notions if its progress is not to suffer permanent arrest. As philosophers it is high time that we inquired into the nature of that which is integral to the very structure of any viewpoint, however abstract or concrete. In short, it is time that we looked to the nature of presupposition.

II

"Philosophy is the search for premises. It is not deduction. Such deductions as occur are for the purpose of testing the starting-points by the evidence of the conclusions." [2]

A presupposition, according to the *American College Dictionary,* is that which is taken for granted in advance, required or implied as an antecedent condition, or supposed as self-evident. More commonly, presupposition has been taken to mean that which is to be assumed without question, simply given. Any tenet, maxim, or postulate satisfying any of these alternative definitions will, for our purposes, be regarded as a presupposition. In the annals of philoso-

---

2 Whitehead, A. N. *Modes of Thought.* New York: Macmillan, 1956. p. 143.

phy many such may be discerned. The history of thought is in no small measure a history of the repudiation of conditions at some time or other supposed self-evident. It is, I think, unnecessary to review the history of intellectual embarrassment.

Presuppositions fall naturally into two types, secondary and primary. A secondary presupposition being one which itself presupposes a primary presupposition. For example, the presupposition of some aestheticians that beauty is proper porportion itself presupposes that nature is, in at least some aspect, uniform. The presuppositions of every branch of philosophy whose topic is less than the nature of ultimate reality are secondary in the sense of implying something antecedent to themselves on which they themselves depend. Precisely the reverse is true of primary presuppositions. That these must be very few in number is obvious. What this exact number is is rather more difficult to determine. On any listing it should, I would say, be no larger than five or six. Leaving aside for the moment such presuppositions as are central to any philosophical theology, four may be distinguished as absolutely fundamental. These basic presuppositions are (1) the uniformity of nature; (2) The univocacy of logical form; (3) the taking of either Being or Becoming as having ontological primacy over its other; and (4) the reality of sense-experience. There is, I think, no major philosophy current today which does not take at least one of these four primary presuppositions for granted.

The authority of Hume and Whitehead seems sufficient evidence for the conclusion that all reasoning from cause and effect takes its charter from the assumption of the uniformity of nature,[3] that on the assumption of this postulate all induction rests.[4] All deduction assumes and must assume some *one* system of logic as its ground. All rational discourse presupposes some specific doctrine of acceptance or rejection of one or more of the traditional laws of thought. There is no metaphysic and no ontology which fails to take its stand either on or between Being and Becoming. The primacy of either, or a perpetual oscillation between the two, is a postulation common throughout the whole history of philosophic systems. Finally, and in many ways perhaps the most important in this age of analysis, is the widespread conviction that any viewpoint not self-restricted to sense-experience and common-sense may be taken as a species of mere linguistic proposal. No other primary

---

3 Hume, David. *An Enquiry Concerning Human Understanding*, Section IV, Part I, Oxford: Clarendon Press, 1946.

4 Whitehead, A. N. *Process and Reality.* New York: Social Science, 1941. pp. 309-313.

presupposition possesses half so much the quality of dogma as that which holds the verifiable in principle or fact to be the sole warrant of truth.

There are two points here that must be noted. The first is that all of the above-mentioned presuppositions may be, and often have been, taken as self-evident. The second, and for our purpose the all-important point, is that *no one of these presuppositions has been or need be accepted without question.* We are confronted with a plurality of presuppositions, for each of which self-evidence may be claimed. We have now to enquire whether or not such self-evidence is really the case, or whether it is perhaps not truer to say that no one of these presuppositions must in itself be taken for granted or assumed as ultimate given fact.

The witness of philosophic history is surely against him who would maintain that it is not only justifiable but essential to take some particular presupposition or presuppositions as self-evident. Each of the four basic presuppositions cited above has at one time or another been brought into question. The postulation of the uniformity of nature is at variance with much contemporary evolution theory. The dependence of the Aristotelian formulation of the laws of thought upon the presupposition of the validity of the substance-attribute mode of predication is well known, and the substance-attribute form of expression itself has been subject to such attack as to render it considerably less than self-evident. The laws of non-contradiction and identity have themselves been subject to a diversity of interpretation, and in the philosophies of the Hegelian tradition their very function has been challenged. The multiplicity of metaphysical interpretations with regard to the ontological status of Being and Becoming is sufficient witness to the inability of any one such view to command necessary assent. The challenge to the self-evidence of that sense-experience which is the sine qua non of contemporary empiricism goes clear back to Plato. There is, in short, no one presupposition or group of presuppositions in all philosophy which may not be or has not been at some time called in question.

Which is, I think, as it should be, for the end result of dogmatism in philosophy can only be dilemma. On the one hand the dogmatist is challenged to find a presuppositionless philosophy and this he cannot do. On the other he is called upon to prove any one of his presuppositions self-evident and this he cannot do either. In no region of philosophic inquiry is he free from presupposition;

in no region can he take any presupposition for granted. Cultural relativists will here remind us that the disjunction offered is not exhaustive, that the dilemma may be avoided by the adoption of that relativism which denies the validity of presupposition altogether. Yet it would seem that even cultural relativists must assume something—even if it be no more than the principle of universal relativity. And if this is so, then it would appear that this minimal assumption of the relativity of any viewpoint whatever is still—assumption. The denial of presupposition itself involves presupposition.

III

"A clash of doctrines is not a disaster—it is an opportunity." [5]

Beginning with the conviction that the nature of the philosophic enterprise involved the autonomous rather than the adjunctival, we found this enterprise, methodologically speaking, taking the form of a critique of presuppositions: exploration of the meaning of this critique, in turn, has led us to conclude that there is no such thing as a presuppositionless philosophy, and no presupposition whose self-evidence is beyond question. If so much be conceded, the strife of systems is not only inevitable but desirable, insofar as it enforces comparison and influences choice among alternatives. What we have now to determine is how such requisite comparison of cosmologies is to be effected.

For comparative cosmology two approaches suggest themselves. We might compare different systems (categoreal schemes) as wholes directly one with another. Or we might, provisionally adopting the standpoint of any one such system, explore its account of itself and its competitors. The first alternative assumes that it is possible to take a judicial position (standpointless) relative to the strife of systems beyond that strife itself. At first glance this would seem to be precisely what we do when we have recourse to the critique of presupposition. Actually, however, this is not so. Indeed, precisely the contrary is the case, as we see when we recall our assertion of the impossibility of a presuppositionless philosophy. Given the truth of such an assertion, standpointlessness is an impossibility. If we compare two standpoints, it must be from some third standpoint that we do so. If we criticize the presuppositions of some one system, or critically compare two or more sets of pre-

---

    5 Whitehead, A. N. *Science and the Modern World*. New York: Macmillan, 1948. p. 266.

suppositions, it must be on the basis of some viewpoint which itself entails presuppositions. Hence only on the second of the approaches noted above does the comparison of cosmologies become a practical possibility.

The first concrete expression, then, of the philosophic enterprise should be the development of a critique of presupposition *within the framework of a categoreal scheme* (cosmology) whose own presuppositions are not themselves exempt from the canons of such a critique. Such a cosmology would be relativistic to the degree that it would account for the developments of past cosmologies and leave open the future development of positions as yet unrecognized. It would be an *open system* in that it would embody some sort of transformation formulae whereby it could account for its own transmutation or transcendence as it confronted the development of the future. It goal would be the ever-increasingly adequate description of the universe within and around us, a goal constantly attained yet never finally realized. Its criterion for choice between descriptions would be pragmatic yet it itself would not be pragmatism. Neither exclusively idealistic or realistic, it would in a sense be both, since for it both change and permanence, Being and Becoming, would be acknowledged to be real factors in fact. Only within such a framework does a critique of presuppositions adequate to its task become a genuine possibility.

Given such a framework, it is possible to derive canons for the critique of any presupposition or set of presuppositions whatever. The number of canons distinguished will naturally vary according to the degree of theoretical precision demanded. Also the canons derived will differ as we are concerned with either primary or secondary presuppositions. However, there are, I suggest, at least four such that must be explicated; the basic canons for the critique of each type of presupposition being two in number.

The first of the canons of secondary presupposition is the canon of experiential conformity, which may be formulated as follows: If the presupposition under consideration be such as to conflict with the testament of experience as received at present,[6] then the validity of the presupposition must be held in doubt pending other justification. Note that I say 'held in doubt' and not 'denied'. It should be plainly understood that the function of the canons is not to serve as a basis for judging of the validity or invalidity of the presuppositi-

---

6 This must be taken to include not merely ordinary sense-experience, but the testimony as to the nature of reality which can be derived from scientific instruments of various sorts, and also the indications as to the character of the real provided by empirically validated scientific theory.

tions, but rather is it their function to render explicit those factors on the basis of which a non-dogmatic judgment may be effected. In sum, the canons are tools for the critique of presupposition, not instruments for its refutation. For example, it is a fundamental presupposition of Kantian epistemology that the pattern and order in nature is imposed by mind, and not (as is commonly held) derivate from nature itself. Applying the canon of experiential conformity to the presupposition, we have to ask whether or not such a presupposition conflicts with experience as presently received. Are we, for instance, justified in assuming nature apart from mind to be an unorganized manifold? Or does an appeal to the witness of the measuring and recording instruments of modern science justify the denial of such assumption? The canon itself provides no final answers to these questions. What it does do is to indicate what questions must be decided before any decision as to the validity or invalidity of this particular presupposition can be properly taken.

The second canon for the critique of secondary presuppositions is that of internal logical consistency: If upon examination any presupposition should prove to be internally and formally inconsistent; that is to say, if the presupposition should be such as to permit the deduction from it of propositions which, *within the framework of the logic in which it is expressed,* are either contradictory or contrary, then the presupposition must be held dubious until justification of it on some other plane is forthcoming. The scholastic presupposition of God as pure act, immutable, omnipotent, and omniscient provides an example in point. The well-known paradoxes of scholasticism may on this analysis be conceived as exemplifying the mutually contradictory propositions deducible from the presupposition in question. For instance, we might cite the paradoxes of good and evil, free-will and determinism, time and eternity, etc., all of which derive from the original presupposition.

When we turn to consider the framing of canons for the critique of primary presuppositions we are at once faced with a rather formidable problem. The canons invoked for the critique of secondary presuppositions are here inapplicable since they themselves imply acceptance of some primary presuppositions. For example, the canon of experiential conformity clearly implies presuppositions as to the uniformity of nature and the reality of sense-experience. Moreover, it would seem that any canons we might derive must themselves be free from the implication of presupposition, must in fact be of a nature more ultimate than that of the primary presuppositions themselves. But that anything more ultimate than these latter is expressible or even possible appears doubtful. Thus it is that any canons framed with regard to the critique of primary presuppositions will by the very nature of the case be radically different in type from those cited above. And this conclusion will hold regardless of whether the canons derived be as we have stated them or not. For the primary presuppositions cannot be evaluated on the basis of experiential conformity or internal logical consistency alone.

There are, however, two methods of approach still open, expressible by what we may call the canons of self-evidence and comparison. According to the canon of self-evidence, if the character of any primary presupposition is such as to permit any reasonable doubt as to its self-evidence, that is, if the denial of it or the assertion of a presupposition contrary or contradictory to it expresses a cosmological possibility, then the presupposition in question cannot be justifiably held more than hypothetical in nature. For example, the presupposition of the uniformity of nature is accepted without question by many philosophers. Can it be doubted? Some will hold that it cannot, arguing that to deny this is to remove the possibility of framing any judgments true for more than the immediate present. Yet on the other hand, the presupposition of the uniformity of nature does not of itself stand as self-evident. There are skeptics who find no long range order whatever in nature. And the position of Gorgias, however much it may offend, is surely not self-contradictory. Then too, we might cite that line of thought common to many philosophies of evolution, which has perhaps been most carefully expressed by Whitehead, *i.e.*, the inevitability of change is such that the laws of nature themselves cannot be held immutable.[7] True, such a view presupposes the future existence of *some* pattern, but if patterns may shift with time, uniformity is hardly self-evident.

Conclusions of a somewhat similar nature may be drawn also from the consideration of the second canon of primary presupposition, the canon of comparison. According to this latter, if a comparison of two or more primary presuppositions with regard to those characteristics differentiating one or the other fails to produce any adequate grounds for either affirming or denying the superiority of one to the other, then the apodictic character of either must be denied, and both will stand, pending other justification, as alternative hypotheses. Applying this canon to our previous example, we institute a comparison of the primary but contradictory presupposi-

---

7 Whitehead, A. N. *Adventures of Ideas.* New York: Macmillan, 1933. p. 143.

tions of the uniformity of nature and the inevitability of change as concerns their relative self-evidence or their adequacy as tenets governing cosmological descriptions. Should this comparison, as is most likely, yield no clear ground for deciding in favor of one or the other of these presuppositions, then we would seem justified in no other conclusion than that each of the presuppositions in question is essentially hypothetical. To adopt either would then be to adopt that one which best satisfies such pragmatic criteria as might be invoked.

To recapitulate: provisionally, we may distinguish four canons for the appraisal of presuppositions. Two of these, the canons of experiential conformity and internal logical consistency, provide us with tools for assessing the worth of secondary presuppositions. Two others, the canons of self-evidence and comparison, can be invoked for the assessment of primary presuppositions. This list of canons is not exhaustive, nor need they be expressed precisely as we have given them.[8] They do, however, constitute a minimal group, lacking any one of which no critique of presupposition is possible.

IV

"In its solitariness the spirit asks, What, in the way of value, is the attainment of life? And it can find no such value till it has merged its individual claim with that of the objective universe. Religion is world-loyalty."[9]

Two viewpoints, hitherto unconsidered, remain to be dealt with before the philosophic enterprise can stand as we have stated it. We will be told by the voluntarist that our conception is vitiated if it be true that whatever presuppositions are invoked by any philosopher are completely arbitrary, and this because philosophy, when concerned with Reality, is purely voluntaristic. And religionists will tell us that we must somehow bring within the domain of our critique those presuppositions involved in the religious experience. Let us consider the problem posed by voluntarism.

If philosophy qua metaphysics is voluntarism, then it would appear that the stress on presupposition as a clue to the comparison of cosmologies is somewhat misplaced. If Will and not Reason is the motivating force, the introduction of a critique of presupposi-

---

8 For instance, in mathematics, the canon of logical consistency finds expression as the consistency of an axiom system (see Wilder, R. L. *Introduction to the Foundations of Mathematics*. New York: Wiley, 1952. pp. 23-27) or a postulate set (see Kattsoff, L. O. *A Philosophy of Mathematics*. Ames: Iowa State College Press, 1948. pp. 236-240).

9 Whitehead, A. N. *Religion in the Making*. Cambridge: Cambridge University Press, 1926. p. 60.

tions can at best be irrelevant, for in this case that philosophy must prevail which is in accord with either the individual or the unversal Will. The nature of the philosophy prevailing will, of course, vary with the variety of voluntarism adopted. But in the end the result will in any case be the same. Insofar as presupposition applies it will do so only within the framework of that thought which is a factor in the realization of Will, and not something for itself. In other words, voluntarism's contention is that if voluntarism is true then the critique of presuppositions has no application.

The fallaciousness of such a mode of argument should be apparent. Had voluntarism universal acceptance the argument might give the appearance of cogency, yet even in such a case it would not be beyond question. For pure voluntarism must assume many factors. It presupposes, for example, that where Will is individual, there the strongest will must prevail, or if Will be regarded as an underlying universal force, it presupposes a harmony of the universal with itself and with all its parts. If the voluntarism held be limited in nature, then the primacy of will over intellect is the basic presupposition. At all events, whether voluntarism is pure or limited it is still a metaphysics. No more than its intellectualistic rivals for philosophical acceptance can it avoid presupposition, no more than they can it claim its presuppositions self-evident.

The case seems somewhat different when we turn to the consideration of religion. Faith itself can hardly be called a presupposition, for reason, according to the religious, cannot be invoked to judge of that whose primary character is the very rejection of reason. Faith is only faith by virtue of its holding to that which is for reason simply absurd. So much can we admit. If the religious viewpoint held be *credo quià absurdum est,* the sole function of the critique of presupposition is to make sure that the viewpoint does not assume a quasi-philosophical guise. For should it do so, should one shift from the *credo quia absurdum est* to the *credo ut intelligam,* then the position held is, regardless of its protestations, but another form of metaphysic, and as such subject to reason's criticisms.

This becomes plain when we look to the notion of God. The ontological argument has for long been recognized by most philosophers as a petitio. In terms of the position here advocated, it presupposes the existence of that which it is its aim to prove. A similar case might be made out with regard to the remainder of the traditional arguments for the existence of God. What, however, is for our purpose of much greater importance, and what needs to be stressed, is that any and all ontology, metaphysics, and theology which maintains that the notion of God is more than hypothetical presupposes as self-evident a notion which the history of the warfare between science, religion, and reason reveals as utterly controversial. If it is impossible to presuppose even the uniformity of nature as final truth, how much less are we entitled to speak with finality about God!

This is not to say that philosophy as comparative cosmology entails acceptance of atheism or agnosticism. It surely does not. What it does require, however, is that the presupposition of the existence and function of God, whether conceived as personal or impersonal, as Absolute or as Force, as Will or as some form of categoreal ultimate, be regarded as being on the same level with all other primary presuppositions,[10] and subject to the same critique as they. In sum, to the extent that a religion employs reason to justify its status as a religion, to that extent are its assumptions subject to the critique of presuppositions. For there is no view, theological, voluntarist, rationalist, or empiricist which is by simple virtue of its subject-matter self-evident or sacrosanct. There is no view which fails to entail presupposition, or whose presuppositions are beyond cavil. It remains that the task of the philosopher is to see the universe as a Whole and to grasp its meaning as a Whole. If this is impossible, philosophy is impossible—but so also are science and religion.

For centuries philosophers have held petitio principi a fallacy. In one sense they have been utterly wrong. Wrong, because every judgment, every chain of reasoning, every system, must eventually beg the question. For as there is no thought without presupposition, so is there no presupposition completely certain. The fallacy lies not in arguing from presuppositions, but in presupposing them beyond the requirement of justification.

Finally, it must be stressed that there is a vast difference between philosophy and *my* philosophy. If philosophy is comparative cosmology, *my* philosophy is simply one view for comparison. Is there, then, one view which, on the basis of this comparison, commands our most serious attention because it most adequately expresses the Whole? For myself, the answer to this question is found in the synthesis of the world-views of Hegel, Bradley, Whitehead, and Berdyaev. But the justification of this point of view is a task for another time.

---

10 In those systems where God is not so much a primary presupposition as a secondary one, i.e., those in which He or It is an exemplification of some perspective or dimension of the system (I am thinking of the function of God in the system of Karl Heim, and to a somewhat lesser degree in the system of Whitehead), the analysis will ,of course, proceed in accord with the canons of secondary presupposition.

TULANE STUDIES IN PHILOSOPHY
VOLUME I, 1952
TABLE OF CONTENTS

*This volume may be purchased for $2.00, plus postage,
From The Tulane University Bookstore,
New Orleans 18, La.*

TULANE STUDIES IN PHILOSOPHY
VOLUME II, 1953
TABLE OF CONTENTS

*This volume may be purchased for $2.00, plus postage,
From The Tulane University Bookstore,
New Orleans 18, La.*

TULANE STUDIES IN PHILOSOPHY
VOLUME III, 1954
TABLE OF CONTENTS

*This volume may be purchased for $2.00, plus postage,
From The Tulane University Bookstore,
New Orleans 18, La.*

TULANE STUDIES IN PHILOSOPHY
VOLUME IV, 1955
TABLE OF CONTENTS

*This volume may be purchased for $2.00, plus postage,
From The Tulane University Bookstore,
New Orleans 18, La.*

TULANE STUDIES IN PHILOSOPHY
VOLUME V, 1956
TABLE OF CONTENTS

This volume may be purchased for $2.00, plus postage,
From The Tulane University Bookstore,
New Orleans 18, La.

TULANE STUDIES IN PHILOSOPHY
VOLUME VI, 1957
TABLE OF CONTENTS

This volume may be purchased for $2.00, plus postage,
From The Tulane University Bookstore,
New Orleans 18, La.